Single-Instruction Multiple-Data Execution

Synthesis Lectures on Computer Architecture

Editor
Margaret Martonosi, *Princeton University*

Synthesis Lectures on Computer Architecture publishes 50- to 100-page publications on topics pertaining to the science and art of designing, analyzing, selecting and interconnecting hardware components to create computers that meet functional, performance and cost goals. The scope will largely follow the purview of premier computer architecture conferences, such as ISCA, HPCA, MICRO, and ASPLOS.

Transactional Memory
James R. Larus and Ravi Rajwar
2006

Quantum Computing for Computer Architects
Tzvetan S. Metodi and Frederic T. Chong
2006

Single-Instruction Multiple-Data Execution
Christopher J. Hughes

ISBN: 978-3-031-00618-0   paperback
ISBN: 978-3-031-01746-9   ebook

DOI 10.1007/978-3-031-01746-9

A Publication in the Springer series
SYNTHESIS LECTURES ON COMPUTER ARCHITECTURE

Lecture #32
Series Editor: Margaret Martonosi, *Princeton University*
Series ISSN
Print 1935-3235   Electronic 1935-3243

Single-Instruction Multiple-Data Execution

Christopher J. Hughes
Intel

SYNTHESIS LECTURES ON COMPUTER ARCHITECTURE #32

ABSTRACT

Having hit power limitations to even more aggressive out-of-order execution in processor cores, many architects in the past decade have turned to single-instruction-multiple-data (SIMD) execution to increase single-threaded performance. SIMD execution, or having a single instruction drive execution of an identical operation on multiple data items, was already well established as a technique to efficiently exploit data parallelism. Furthermore, support for it was already included in many commodity processors. However, in the past decade, SIMD execution has seen a dramatic increase in the set of applications using it, which has motivated big improvements in hardware support in mainstream microprocessors.

The easiest way to provide a big performance boost to SIMD hardware is to make it wider—i.e., increase the number of data items hardware operates on simultaneously. Indeed, microprocessor vendors have done this. However, as we exploit more data parallelism in applications, certain challenges can negatively impact performance. In particular, conditional execution, non-contiguous memory accesses, and the presence of some dependences across data items are key roadblocks to achieving peak performance with SIMD execution.

This book first describes data parallelism, and why it is so common in popular applications. We then describe SIMD execution, and explain where its performance and energy benefits come from compared to other techniques to exploit parallelism. Finally, we describe SIMD hardware support in current commodity microprocessors. This includes both expected design tradeoffs, as well as unexpected ones, as we work to overcome challenges encountered when trying to map real software to SIMD execution.

KEYWORDS

SIMD, vector processor, data parallelism, autovectorization, control divergence, vector masks, unaligned accesses, non-contiguous accesses, gather/scatter, horizontal operations, vector reductions, shuffle, permute, conflict detection

Contents

Preface

The mid-1990s saw personal computers used increasingly for multimedia applications, such as playing audio and video clips. Mass-market microprocessor vendors were motivated to provide a big boost to the performance of these applications, to differentiate from the competition. To that end, Sun Microsystems introduced the Visual Instruction Set (VIS) in 1995, and Intel quickly followed with the highly visible MMX™technology in 1996 (many of us still recall the advertisements with the "bunny suits"). Both vendors enhanced existing instruction sets and microarchitectures with single-instruction-multiple-data (SIMD) support—a single VIS or MMX instruction could operate on multiple small integers, such as color information for pixels in a video. While SIMD was not a new idea, it had previously been confined to high-performance computing, and was mostly used in vector machines. Soon afterward, IBM and Motorola went a step further by introducing the much more powerful AltiVec instruction set—this included support for floating-point, which opened the door to using SIMD on a much broader variety of applications. AMD helped push x86 in the same direction with 3DNow!, an extension to MMX with single-precision floating point support; Intel responded with Streaming SIMD Extensions (SSE).

At the same time, the focus on uniprocessor performance was shifting to clock speed. In the mid-to-late 1990s clock frequency increases in general-purpose processors accelerated dramatically. Manufacturers of desktop processors often competed directly on clock frequency rather than actual performance. For programmers, this was fantastic—getting much better performance automatically across generations meant less effort was needed on code optimization. However, this was short-lived. In the early 2000s, this clock frequency war ended rather abruptly as processor designers were reined in by physical constraints—clock frequency increases had drastically increased processor power consumption, and this was stressing cooling technology, not to mention battery life of the increasingly important laptops.

Despite serious concerns about the readiness of the software ecosystem, processor vendors turned to multi-core designs as a way to increase performance of a single processor package and stay within power constraints. The multi-core era had arrived. Multi-core processors, however, do not provide an increase in single-threaded performance across generations, unless the core design is improved or clock speeds go up. Since clock frequencies have stagnated (or even gone down, to make power headroom for more cores), one might expect that single-thread performance has not improved in recent years. Not so. Architects have continued to improve single-thread performance, to great effect. For example, by some measure, single-thread performance of an Intel Core i7-4790 at 3.6 GHz is more than three times that of an Intel Pentium 4 also at 3.6 GHz [PassMark, 2014]. Some of that performance boost comes from an improved memory subsystem—e.g., bigger caches and faster DRAM. Some of it comes from increasing the number of in-flight instructions that a core can support via a larger reorder buffer and other microarchitectural structures. However, a very large part of the recent boost in single-thread performance comes from SIMD.

As improving single-thread performance within a power budget became increasingly challenging, architects looked to SIMD as a relatively efficient way to greatly boost performance. For applications using SIMD, doubling the SIMD width (i.e., the number of bits in a SIMD operation) can theoretically double performance. Moreover, doubling the SIMD width increases the area and power of a core by much less than a factor of two.

Providing hardware support for SIMD, or increasing SIMD width in processors that already have some such hardware support, does not automatically lead to improved performance—applications must use the SIMD support to realize any benefits. Happily, software has evolved to take advantage of SIMD, for at least two reasons.

First, great effort was put into the software ecosystem to help key software vendors make use of SIMD, and to enable compilers to automatically emit SIMD instructions. In general, features that provide a performance or efficiency boost only with significant software changes suffer from the chicken-and-egg problem: it is hard to justify the hardware support for such a feature since software that can use it doesn't exist, and yet software vendors are not motivated to change their software for unproven hardware features. In the case of SIMD, some version of the hardware was already present in most processors. Further, certain processor vendors mitigated the risk for major software developers by providing direct assistance to add SIMD to their key routines.

Second, the renewed interest in SIMD coincided with the multi-core era. For individual applications, as opposed to multiprogrammed workloads, multi-core processors provide benefits only when the application is parallelized via threading or some equivalent (e.g., message passing). In the server space, multi-socket machines already had this property, so multi-core processors were nothing fundamentally new. However, multi-core processors did lead to a large increase in the thread-level parallelism that an application could exploit, which encouraged algorithm and application developers to better explore that direction. While SIMD execution generally exploits data parallelism rather than thread-level parallelism, for certain algorithms, these are very closely related. Thus, for some key applications, getting software developers to think about parallelism led to implementations that could better utilize both multiple cores and SIMD.

Today, the vast majority of microprocessors include SIMD support. This includes not only processors in high-end servers, desktops, and laptops, but mobile phones, consumer electronic devices, and graphics processors (GPUs). Another synthesis lecture covers general-purpose processing on GPUs (a.k.a. GPGPU), so we will focus on SIMD for CPUs; that said, many of the principles discussed here apply to SIMD on GPUs. This book attempts to summarize for computer architects the concepts of SIMD execution, the costs and benefits of providing various features for SIMD execution, and how SIMD instruction sets have recently evolved to cover challenging computational patterns.

OUTLINE OF THE BOOK

In Chapter 1, we introduce data parallelism, including how it manifests in real-world applications. We provide some specific real-world examples from the scientific and enterprise spaces.

Chapter 2 describes the principles of SIMD, and how SIMD can be an efficient means of exploiting data parallelism. We compare the SIMD approach to alternatives such as superscalar execution. We also discuss SIMD programming models and compilation.

Chapter 3 discusses SIMD computation and control flow. We cover the basics of SIMD arithmetic, more complex computations, and various options for handling control flow.

Chapter 4 discusses SIMD memory operations. We explore how SIMD execution places a lot of pressure on memory systems, especially for indirect accesses, and introduces new challenges like alignment.

Chapter 5 discusses horizontal operations in SIMD, which enable cross-SIMD-element communication. These are key to applying SIMD to non-trivially-parallelizable problems.

Chapter 6 concludes with some thoughts on the future of SIMD execution.

Christopher J. Hughes
May 2015

Acknowledgments

I'd like to thank several people for their help in writing this lecture. Mike Morgan and Margaret Martonosi, for giving me the opportunity and encouragement to put my thoughts on this subject to paper. Chris Batten, for his incredibly detailed and thorough suggestions, to which I tried to do justice on a tight timeline. Pradeep Dubey, for his encouragement and feedback. My current and former colleagues at Intel, who taught me much of what I know on this subject. Finally, my wife, Maureen, for her patience and support.

Christopher J. Hughes
May 2015

CHAPTER 1

Data Parallelism

1.1 DATA PARALLELISM

When mapping a problem to parallel hardware, algorithm designers and application developers focus on the computation they are trying to perform, and on what data. They typically reason about either *functional parallelism* or *data parallelism*—i.e., respectively, performing multiple operations in parallel, or operating on multiple data items in parallel. In most respects, these are equivalent; our choice of term speaks mostly to how we think about the parallelism in an application, and especially to how we exploit it.

When reasoning about data parallelism, we most often think about performing the *same computation* on a set of data items. Parallelization doesn't get much simpler than "we need to do operation X on data items a through z, and this can all happen simultaneously and in any order." Furthermore, this situation arises extraordinarily often in real applications, especially in ones that involve processing or modeling something from the real world. For instance, multimedia data, sensor data of all kinds, and 3-D models are just a few examples of input data sets that can be both large and homogeneous. Many algorithms used to process these data sets treat the elements as independent, and thus, they have abundant parallelism.

Data parallelism can exist even when we perform different operations on different data elements. Otherwise, we would preclude all computations with input-dependent control flow from having data parallelism. We will thus use a common and reasonably general definition of a computation with data parallelism: a computation on a set of independent data items, where we have a single code base and a common starting point within that code for each data item.

In the rest of this chapter, we provide some concrete examples of data parallelism in a wide range of real-world applications. While different algorithms can exhibit widely varying parallelism characteristics, one important pattern that will show itself through these examples is *workload convergence* [Chen et al., 2008]. Workload convergence means that both within a single domain and across multiple domains, we find a relatively small set of primitive operations at the heart of many key algorithms. As a consequence of this, when one of these primitives has a certain property, such as significant data parallelism, many real-world applications will exhibit it, and we can exploit it using the same fundamental technique.

1.2 DATA PARALLELISM IN APPLICATIONS

The following is a small set of example real-world applications with inherently large amounts of data parallelism. Exploiting this parallelism can sometimes be challenging, and as we describe the applications, we will point out some specific instances of challenges that we tackle later in the lecture.

1.2.1   PHYSICAL SIMULATION

Physical simulation is a broad field, since scientists, movie animators, and others are interested in simulating many different kinds of physical phenomena, such as fluids, cloth, hair, muscles, objects fracturing, earthquakes, and many more. We briefly describe just a couple specific examples.

Computational fluid dynamics (CFD) applications simulate the movement of fluids, including interaction with solid objects. The simulation technique often depends on both the required fidelity of the simulation (e.g., scientific simulation vs. "Hollywood" simulation vs. real-time game simulation) and the environment (e.g., water in a confined space like a glass, an ocean, and/or whether the fluid interacts with solid objects). Most fluid simulation applications attempt to solve, to some level of precision and under some set of assumptions, the Navier-Stokes equations. These equations describe the velocity field of a fluid at a given time, or the direction and speed of the fluid at every point in space at that time, as a function of the pressure of the fluid and the gradient of the velocity field. A popular method to solve the equations is to first discretize the space being simulated, i.e., divide the volume into a 3-D grid of voxels. We then apply a finite difference method [Foster and Metaxas, 1996], which essentially updates each voxel using values in the neighboring voxels. This may sound daunting, but computationally, this is fairly straightforward. Essentially, we iterate over time, and at each time step, we perform a computation over all voxels; the new velocity at a voxel is a function of the pressure and old velocities of the neighboring voxels, including the faces and edges of the voxels, as shown in Figure 1.1. This is an example of a *stencil computation*, a key primitive in many algorithms.

Stencil computations naturally arise in many places, including applications that solve a set of partial differential equations, like CFD applications. While there are many methods for doing this, one of the most common is discretization of the problem space and applying a stencil computation across all grid cells in parallel. Thus, applications that include solving partial differential equations often exhibit plenty of data parallelism, and furthermore, they can often borrow the same techniques for exploiting it. Since sets of partial differential equations describe numerous real-world problems, these techniques are applicable in myriad domains beyond fluid simulation, such as image processing and financial analytics.

Another type of physical simulation application, a molecular dynamics application, simulates the movement and interaction of particles at the molecular level, and is used by physicists, chemists, and biologists alike. Scientists are interested in a range of molecular interactions, and each involves different force computations between neighboring particles. However, the simulations share a common framework in that, for each time step of the simulation, they compute the total force exerted on each particle by all the other particles, use that to update the particle's velocity, and use that to update the particle's position. As shown in Figure 1.2, the force, velocity, and position computations are all data parallel; the force computation is typically the most expen-

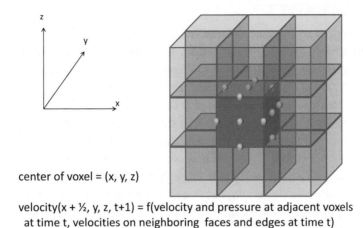

center of voxel = (x, y, z)

velocity(x + ½, y, z, t+1) = f(velocity and pressure at adjacent voxels
    at time t, velocities on neighboring faces and edges at time t)

Figure 1.1: Data parallel velocity computation for fluid simulation. This computation may be done for all voxels simultaneously.

sive by far [Pennycook et al., 2013]. One challenge presented by molecular dynamics applications is that the particles being simulated are moving around, and thus gain and lose neighbors over time. Thus, before each time step, we cannot know exactly which particles another will interact with, for distance-dependent computations. This introduces an input-dependence on the amount of computation we do for each particle. It also means we may have different control flow for different particles.

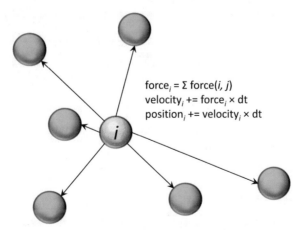

$$\text{force}_i = \Sigma \text{ force}(i, j)$$
$$\text{velocity}_i \mathrel{+}= \text{force}_i \times dt$$
$$\text{position}_i \mathrel{+}= \text{velocity}_i \times dt$$

Figure 1.2: Data parallel force computation for molecular dynamics. This computation may be done for all particles simultaneously.

1.2.2   COMPUTER VISION

Computer vision applications, such as face detection, autonomous driving, and anomaly detection in security camera feeds, typically consist of a mixture of image processing and (specialized) machine learning algorithms. One application of recent interest is articulated body tracking [Chen et al., 2007]. Here, using one or more video feeds, a computer detects that a person is in view, and estimates the location and position of each of his/her limbs. The application begins processing each video frame with an image processing step, e.g., by denoising it via a median filter (see Figure 1.3). A median filter is another stencil computation, where each pixel is recomputed as the median value of all pixels in its neighborhood. One challenge presented by this type of computation, and indeed all stencil computations, is how to handle the boundaries of the image—boundary pixels have too few neighbors in one or more directions to "fill" the stencil. Therefore, these pixels need different handling than others, giving us nonuniform control flow. The next steps in the application are foreground detection and edge detection, which, respectively, determine which pixels are part of the foreground and which pixels are on edges in the image; these are similarly data parallel because all pixels can be processed concurrently. Finally, candidate poses (possible positions of a person and his/her limbs) are tested by checking how close of a match they would have with the detected foreground and edge pixels; many poses are tested, and all may be done concurrently.

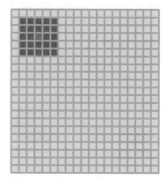

New color of blue pixel =
Median color value of red
and blue pixels

Figure 1.3: Median filter used in image processing and computer vision applications. This computation may be done for all pixels simultaneously.

1.2.3   SPEECH RECOGNITION

Speech recognition applications, such as those used by mobile phones as an alternative input method, contain two primary stages: extracting features out of each snippet (e.g., 10ms long piece) of raw audio input, and inferring the most likely speech based on the current and previous snippets. Gaussian mixture models are often used for feature extraction [Stuttle, 2003]. These convert each snippet of sound to a set of probabilities, i.e., for each one of a set of pre-determined symbols, they compute the likelihood that the snippet contains that symbol. The probabilities can be computed concurrently. For the inference engine, many implementations today use a weighted finite state transducer (WFST) [Chong et al., 2009]. A WFST is a state machine that incorporates an acoustic model, pronunciation model, and language model. Each state change has an inherent likelihood (e.g., based on the language), may consume a specific symbol, and may output a specific word. Since state changes are probabilistic, we track a number of most likely current states (e.g., thousands or tens of thousands), called *active states*, and their most likely path from the starting state; the path corresponds to a candidate string of words. As each new sound snippet is processed, we use it to consider state changes for each of the active states. This data parallel operation is shown in Figure 1.4. Parallel implementations of WFST, and other graph algorithms that update properties of the graph, must deal with the possibility of conflicting updates. For instance, we must prevent two workers from attempting to update the same state in the graph simultaneously, or somehow combine their updates.

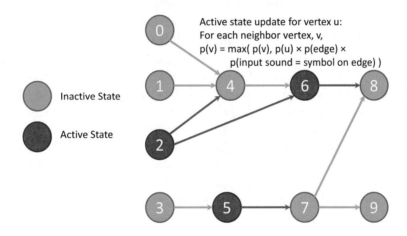

Figure 1.4: WFST during an active state update. All active states may be updated simultaneously.

1.2.4 DATABASE MANAGEMENT SYSTEMS

Database management systems (DBMSs), in part, process user queries on a collection of data. Parallelism may exist between queries that access disjoint pieces of the database, or have non-conflicting (e.g., read-only) accesses to overlapping parts of the database. We may view this as data parallelism, especially if queries are batched in some manner. There is also often data parallelism within a single query, as we apply a common operator to multiple data items. One concrete example is sorting, a particularly important operation for DBMSs. A DBMS may use sort because a query directly requests it, or for other reasons, such as join operations. Merge sort and radix sort are popular choices for sort algorithms in DBMSs, and both have significant data parallelism [Satish et al., 2010]. Merge sort's data parallelism is in the comparisons between elements; many independent comparisons are done at each step of the algorithm. Radix sort's data parallelism is in computing the radix for each input value, and also in writing each input value to its correct position in the output list. Determining the correct position for each input value does require communication and/or synchronization between workers, making this challenging to parallelize efficiently.

Index searching is another important operation in DBMSs with similar data parallelism. The fastest index searches are done on binary trees [Kim et al., 2010]. In addition to data parallelism across independent searches, each search key may be compared to multiple indices in the tree simultaneously.

1.2.5   FINANCIAL ANALYTICS

Computational finance includes applications such as risk management and portfolio analysis, but perhaps the most visible work in this area is on derivative pricing—if we can accurately determine how much an asset is worth, we can judge whether we should buy or sell it at a given price. Much effort has gone into pricing options, in particular, including the development and optimization of several different techniques [Smelyanskiy et al., 2012]. These include the well-known Black-Scholes equation, which can be solved analytically. Black-Scholes implementations typically try to price large numbers of options, and thus exhibit data parallelism across the pricing of different options. American options are harder to price than European ones, since American ones can be exercised before they expire. Thus, more expensive techniques are used for that, such as Crank-Nicolson. Crank-Nicolson uses finite differences, as we described for fluid simulation earlier. However, the lattice we form here is only two dimensional, and instead of representing anything physical, has time on one axis and asset price on the other. As with fluid simulation, we compute values at each point on the lattice based on the values of neighboring points. Data parallelism exists between different points in the lattice, as with fluid simulation, and also between different options, as with Black-Scholes.

1.2.6   MEDICAL IMAGING

Medical imaging applications process the results of scans such as X-ray radiographs and magnetic resonance images to reconstruct a visual representation of internal tissue and organs, often in three dimensions. Given a 3-D representation of tissue and organs, one common technique for rendering 2-D images from it, e.g., for a physician to perform a diagnosis, is ray casting [Smelyanskiy et al., 2009]. Ray casting traces rays from the observer's eye through each of the pixels of the output image, as shown in Figure 1.5. To determine the final color of each pixel, it accumulates the color contribution of every voxel the ray passes through, which depends on the contents of each voxel. Since the rays are independent, this computation is data parallel.

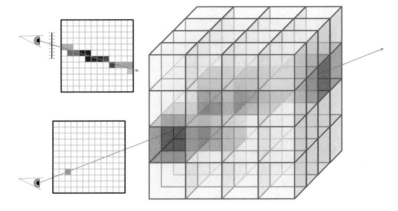

Figure 1.5: Ray casting. Top left: Casting a ray from an observer, through a 1-D output image, and through a 2-D scene. Bottom right: Casting a ray from an observer, through a 2-D output image, and through a 3-D volume. The cells that a ray hits (lit up here) all contribute to the final color of the pixel. We may trace rays for all output pixels simultaneously.

CHAPTER 2

Exploiting Data Parallelism with SIMD Execution

2.1 EXPLOITING DATA PARALLELISM

As architects, we have a few options for increasing performance of a single processor core.

1. Decrease the time needed to execute each instruction by increasing processor and/or memory system frequency, or via microarchitectural optimizations such as reducing the latency of instructions or memory

2. Execute more instructions simultaneously, e.g., via wider superscalar execution or by increasing the number of threads running concurrently

3. Increase the amount of (useful) work each instruction does, so a given task requires fewer instructions

Executing each instruction faster accelerates programs of all kinds, but is typically done primarily to speed up serial execution rather than parallel execution.

In contrast, executing more instructions concurrently is all about exploiting parallelism in the instruction stream, including data parallelism. For example, data parallelism can manifest as multiple independent, nearby instructions in a single thread—a core can potentially execute them simultaneously. Increasing instruction concurrency is quite general, though, and does not specifically target data parallelism.

The last option, making each instruction do more work, is an idea rooted in complex instruction set computing (CISC). CISC instructions combine operations that are often done together, to allow hardware to perform the combined operation more efficiently. For example, a CISC instruction might combine a load, arithmetic operation on the loaded value, and a store of the result. This combination means the instruction fetch and decode units do less work than if each operation were separate instructions. When we combine dependent operations into one instruction, as in our example, we may also avoid writing intermediate results to registers.

In this book, we focus on a combination of the second and third options, single-instruction multiple-data (SIMD) execution. SIMD borrows the CISC concept of an instruction that specifies multiple operations, but to combine *independent* operations. Further, the independent operations are the same arithmetic function, but on different data elements. That is, SIMD instructions tell the hardware to apply the same operation to a *set* of independent data elements. Thus, SIMD specifically targets data parallelism.

2.2 SIMD EXECUTION

SIMD is aptly named—a **single instruction** tells the hardware to perform a given operation on **multiple data** elements. This explicitly exposes parallelism to the hardware—the processor knows that the operations specified by the instruction can be done simultaneously. Architects can leverage this to increase the performance and/or energy efficiency of a core, as we will explain in Section 2.3

Supercomputers were the first to adopt SIMD execution. Illiac IV was the first machine to use SIMD execution [Barnes et al., 1968], and was followed by the CDC STAR-100 [Hintz and Tate, 1972] and Texas Instruments ASC [Watson, 1972]. The latter two more specifically contained the first *vector processors*. A vector processor operates on groups of independent data elements, i.e., *vectors*. For example, they may specify with one instruction that 64 elements of array A should be added to 64 elements of array B; the underlying hardware performs one or more additions per cycle, in a pipelined fashion, until it completes all 64 additions. The operands are typically stored in vector registers, registers capable of holding an entire vector.

The original vector processors and some more recent instantiations used deeply pipelined functional units to allow for high frequency operation. The drawback to this is that the vectors need to be relatively long, to cover the pipeline depth. As we discussed in Section 1.2, many computations have plenty of data parallelism, and can use long vectors. However, most of these computations also require a certain amount of short vector operations or even scalar operations (i.e., operations on single data items). Amdhal's Law rears its head here, and these short vector or scalar operations limit the performance of vector processors that rely too heavily on pipelining.

Further, applications run on mass market hardware (i.e., desktop and laptop computers, tablets, and smart phones), typically have considerably less data parallelism at each step of an algorithm than high-performance computing applications. Even when running applications dominated by data parallel computations, the problem sizes in those environments are smaller. Long vectors are therefore not the best match.

Instead, today's general-purpose processor vendors generally exploit SIMD through support for short vectors. Rather than allowing for vectors of 64 or more double-precision elements, they have maximum vector lengths of between two and eight double-precision elements. Another key difference from the original vector processors is that instead of using pipelined functional units to process a single vector operation, most of today's SIMD implementations instead use wide functional units. These functional units may be 128-bits, 256-bits, or even 512-bits wide, and perform multiple, independent additions, multiplications, etc., simultaneously. Processors generally support SIMD operations on both integer and floating-point data types of various widths. For example, a 128-bit vector may comprise two double-precision float-point elements, or it may comprise 16 8-bit integers. The functional units themselves are often still pipelined to allow higher frequency operation, but in most cases, an entire SIMD register worth of data is processed by a functional unit in one cycle. There are some exceptions, where the logical SIMD width of the processor (i.e., the number of bits specified within an instruction) is larger than the functional unit width. In these cases, the processor relies on a shallow pipeline (e.g., two or four cycles deep) to complete execution of one SIMD instruction.

Modern SIMD memory systems operate similarly to SIMD arithmetic logic units (ALUs)—they load or store multiple elements simultaneously. Since caches store data in contiguous chunks, SIMD loads and stores are most efficient when accessing contiguous memory locations. In such cases, they read or write a contiguous part of a cache line, or one entire cache line. Non-contiguous memory operations are significantly more complicated, since they frequently require accesses to multiple cache lines. Many processors do not even support non-contiguous SIMD memory operations, given the implementation challenge, and the generally low performance of such operations.

Given the differences between the original vector processors targeting long vectors, and their modern, short vector counterparts, a terminology distinction is often made between the two. When most architects discuss SIMD, they refer to the short vector variety; the term vector processing is typically reserved for processors supporting long vectors, and often assumes pipelining of vector elements. This lecture concentrates on the more widespread instantiation of SIMD, support for short vectors. However, most of the concepts apply equally to long vector execution. We therefore make no terminology distinction between short and long vectors; this lecture uses the terms "SIMD execution" and "vector execution" interchangeably.

2.3   SIMD PERFORMANCE AND ENERGY BENEFITS

Processors employ SIMD execution to quickly and efficiently execute data parallel operations. A SIMD instruction has two properties that translate to advantages in various parts of the microarchitecture: (1) it combines multiple operations into a single, compact instruction, and (2) it explicitly indicates that the operations comprising it are independent.

Example 2.1   Czechowski et al. [2014] provides a concrete example of the performance and energy benefits from SIMD execution. That study compared the performance and power consumption of an Intel Haswell processor running the Livermore Loops with 128-bit wide and 256-bit wide SIMD. For a loop designed to stress floating-point throughput, the wider SIMD doubles performance with only a 5% increase in power consumption, providing a 1.9x improvement in energy efficiency. Other loops with less data parallelism see smaller benefits from wider SIMD, but energy efficiency still improves by 4–56%.

As outlined in Section 2.1, there are alternative, more general techniques for exploiting data parallelism. We now compare SIMD with the most common of these other techniques, in the context of exploiting data parallelism. In particular, we compare against (1) a superscalar core that can execute multiple, scalar instructions from a single thread each cycle, (2) a multithreaded core that can execute a single scalar instruction from each of a set of threads each cycle, and (3) a multi-core processor composed of multiple scalar cores. For (1), we do not specify whether the core supports out-of-order execution as it is not relevant to this discussion. For (2), we do *not* add multithreading to a superscalar core, which would combine multiple techniques, but rather consider a core with multiple thread contexts that share a front end and memory system, and where we can execute one instruction per thread each cycle. Sharing execution units among threads to increase utilization and/or peak throughput for a thread can be done in either a scalar context [Smith, 1981] or a superscalar one [Tullsen et al., 1995] However, our goal here is to compare the costs and benefits of specific isolated techniques.

Table 2.1 lists the hardware behavior of various parts of the microarchitecture when exploiting data parallelism, for these different architectural techniques. Figures 2.1 and 2.2 illustrate this, with a simplified version of the designs. For this discussion, we compare designs with the same peak performance.

When fetching and decoding instructions, SIMD designs need only handle one instruction at a time to bring in several operations' worth of work into the core. To match the throughput of SIMD designs, superscalar and multithreaded cores must fetch and decode multiple instructions per cycle; this consumes considerably more area and power, and complicates the design. The fetch and decode part of each core in multi-core processors can be as simple, or even simpler, than in SIMD cores, but it must be replicated in all cores. Thus, SIMD designs have significantly lower cost in terms of area and power than the other designs. The tradeoff is that the SIMD design only achieves high throughput for SIMD instructions—code not dominated by data parallelism will see much lower instruction throughput. The other designs are more general, exploiting parallelism either between instructions within a thread (i.e., superscalar) or across different threads (i.e., multithreaded and multi-core).

SIMD execution constrains all elements within a vector to follow the same control flow path. Some designs include support for predication, which allows a SIMD instruction to conditionally operate on each element. However, with predication the hardware still executes every instruction—some results are simply gated, or thrown away (more discussion in Chapter 3). This inflexible design is simple and cheap. In contrast, the other designs execute scalar operations exclusively, allowing each data element to potentially follow a unique control flow path. For superscalar designs, the high-speed processing of a single thread means that they must predict branches, possibly even multiple per cycle, and that an incorrect prediction will likely carry a large penalty. Multithreaded designs and multiple scalar cores process each thread more slowly than a superscalar design, and so while they also rely on branch prediction, they have smaller penalties for misspeculation.

SIMD designs do not need to check for dependences between data elements in a single instruction, since the instruction guarantees their independence. Dependence checks are needed across instructions (e.g., to decide whether an instruction must wait for another to finish), since one SIMD instruction may consume the result produced in another. However, this check is the

Table 2.1: Hardware behavior when exploiting data parallelism for some common architectures

Hardware	SIMD	Superscalar	Multithreading	Multi-core
Fetch/Decode	Single instruction specifies many instances of same operation	Handle multiple instruction per cycle	Handle multiple instruction per cycle	Each core has own fetch/decode logic
Control Flow	Same code path for many elements, predication	Each element has independent control flow, prediction may be hard	Each element has independent control flow	Each core has independent control flow
Inter-Element Dependence Check	No needed, only check between instructions	Check all instructions with each other	Intra-thread checks, but no inter-thread checks	Intra-thread checks, but no cross-core checks
ALUs	Wide ALU, same operation on multiple elements per cycle	Multiple independent ALUs	Multiple independent ALUs	Each core has own ALUs
Memory System	Wide memory operations, limited non-contiguous support	Multiple narrow operations	Multiple narrow operations	Narrow operation(s) per core, coherence actions

same as that between two scalar instructions, since we simply compare destination and source registers. Multithreaded and multi-core designs, like SIMD designs, must check for dependences between instructions in the same thread, but inter-thread communication is (typically) only supported through memory. Superscalar designs are much more expensive—to support high throughput execution, they must support multiple instructions passing a dependence check every cycle. In addition to performing multiple checks against the already in-flight instructions, these designs must also check the candidate instructions against each other. This extra cost is necessary to support simultaneous execution of multiple instructions from the same thread; this is generally

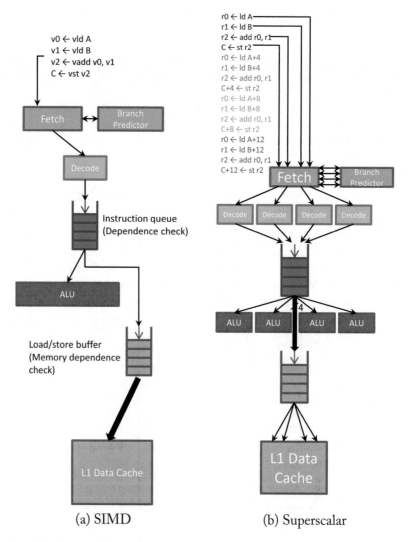

(a) SIMD   (b) Superscalar

Figure 2.1: Simplified view of two processor designs that exploit data parallelism. Each executes part of a computation that conditionally adds a constant to each value in array A.

a limiting factor in superscalar throughput, as it grows quadratically with the issue width of the core [Palacharla et al., 1997].

SIMD ALUs are large and expensive compared to scalar ALUs, since SIMD ALUs process multiple data items simultaneously. However, they are much cheaper than iso-performance alternatives that use multiple scalar ALUs. A set of independent, scalar ALUs, as used by superscalar, multithreaded, and multi-core designs requires separate control logic and input and output data

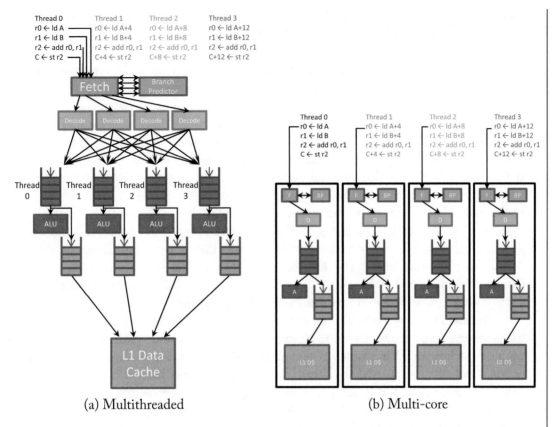

(a) Multithreaded   (b) Multi-core

Figure 2.2: Simplified view of two additional processor designs that exploit data parallelism.

paths. Unlike a SIMD ALU, this allows them all to perform completely different operations, and on different inputs and outputs.

SIMD memory systems support wide memory operations, and may support non-contiguous memory operations. Wide memory operations (e.g., reading a 32B chunk of data) require some additional support in the memory dependence hardware to handle a wider range of load and store widths, and may require extra hardware to detect and buffer data if a single operation touches multiple cache lines. Non-contiguous SIMD memory operations access potentially many cache lines in no particular pattern. To process these at the same throughput as contiguous memory operations, designs require extra address generation and translation hardware, and possibly an enhanced cache to provide high throughput for non-contiguous accesses (Chapter 4 discusses this more). In practice, designs that support non-contiguous SIMD memory operations often do so with a throughput of only a small number of data elements per cycle. To match the throughput of SIMD memory systems, superscalar and multithreaded designs must support issuing multiple, independent loads and stores each cycle. This is essentially the same as supporting high throughput non-contiguous SIMD memory operations, but even more expensive since superscalar and multithreaded designs must also check for dependences between the simultaneously issued memory operations. Multi-core designs can be much simpler, by supporting the issuing of only one (or a small number) of narrow loads and stores each cycle. This hardware, including the data cache itself, must be replicated across each core, however. Further, if required by the programming model, multi-core designs also need hardware for cache coherence.

You might have noticed a consistent theme in the comparison between SIMD and other techniques for exploiting parallelism: SIMD designs are very efficient at exploiting data parallelism in terms of area, power, and design complexity, but this efficiency comes at the price of specialization. Multithreaded and multi-core designs have higher cost since they rely on replicated hardware for each thread, but they can exploit parallelism across those threads even if they perform completely different operations from completely different pieces of code. Superscalar designs are higher cost still. At each stage of the microarchitecture, they require replicated hardware (e.g., ALUs) and/or hardware whose cost scales superlinearly with the instruction throughput (e.g., instruction fetch and dependence checking). The payoff is that they are even more general than multithreaded and multi-core designs; they can extract parallelism across instructions from a single thread, regardless of what those instructions are doing.

In practice, designs typically use some combination of the above techniques. In the server space, for example, the most common processor line is the Intel Xeon processor, which uses *all* of the above techniques. Why? Superscalar cores are great for software developers since they tend to reason about and design serial algorithms, but superscalar cores' limited instruction throughput means we can't rely on this technique alone for high performance designs. SIMD is an efficient way to exploit data parallelism, and can be incorporated into a superscalar core to allow it to achieve higher throughput without stressing the most expensive hardware components. As we will discuss more in the next section, though, there are limits to how large we can scale vectors. Multithreading can help us improve the utilization of superscalar cores [Tullsen et al., 1995], even those with SIMD support, but as the number of threads on a core grows, we place a lot of pressure on a core's memory system, especially when considering the capacity of a first-level data cache. Finally, using multiple cores provides us with a very scalable way to increase instruction throughput of a whole processor, but its inherent replication of everything in the core and reliance on software to expose parallel work as separate threads means that we use this as a method of last resort—designers tend to only add cores once they have a core that they believe hits the sweet spot of performance and cost.

2.4 LIMITS TO SIMD SCALING

What are the limits to the performance and efficiency benefits of SIMD? Why not build a processor with as wide SIMD execution as we have transistors to afford? The answers lie in both physical design constraints and algorithmic limits.

As we just touched on, there are design tradeoffs for any architectural technique for exploiting parallelism. For SIMD, many of the physical costs are independent of SIMD width, such as the area and power cost of instruction decoding. Some of the costs scale linearly with SIMD width, such as the area and power of the execution units. However, the interface to the memory system becomes increasingly complex as SIMD width grows.

Assuming the processor has data caches, as SIMD width grows, the number of cache lines touched by a single SIMD memory operation also grows. Conventional caches access only a single cache line per cache port, and so support only a small number of simultaneous cache line accesses. One possible response to larger SIMD width is to grow the cache line size, but this carries drawbacks. In particular, growing the line size may not reduce the number of lines touched for non-contiguous accesses. Indeed, for non-contiguous accesses, growing the line size *decreases* bandwidth efficiency since we will likely touch less of each line before it is evicted. An alternative is to increase the number of cache ports; this will increase memory throughput for both contiguous and non-contiguous accesses. However, the area, power, and latency costs of adding cache ports grows quickly. Techniques like banking can increase cache bandwidth, but only for non-contiguous accesses, and they have physical limits on scaling [Juan et al., 1997, Rivers et al., 1997].

Similarly, if we grow the SIMD width enough, we will increase the number of pages that each SIMD operation touches. Assuming the processor uses a translation lookaside buffer (TLB) to translate virtual address to physical ones, we face a similar dilemma as we do with data caches: to provide sufficient throughput, we can increase page size to keep the number of pages touched by a SIMD access constant, replicate the TLB and have different vector elements access different copies [Espasa et al., 2002], or increase the throughput of a single TLB, which scales non-linearly [Austin and Sohi, 1996].

Despite these physical design constraints to growing SIMD width, real designs are constrained far more by algorithmic limits. At the highest level of abstraction, we should see performance proportional to SIMD width until the vector length (i.e., SIMD width, in number of elements) exceeds the available data parallelism. Since SIMD execution is about performing the same operation on a set of data items, this means we are limited by the number of data items we process at each step of an algorithm. Real algorithms often include a significant number of scalar steps, or steps with significantly fewer inputs than others. This introduces an Amdahl's Law effect, which limits SIMD efficiency.

Definition 2.2   We define *SIMD efficiency* to be the ratio of the instruction count reduction in the SIMD version of the code to the vector length, as shown in Equation 2.1. For instance, if the vectorized version of a program has one quarter of the instructions of the scalar version, then the instruction reduction is four. If the ISA for the program supports a vector length of four data elements, then the program's SIMD efficiency is 100%. Note that, if the SIMD versions of instructions have the same latency and throughput as their scalar counterparts, then the instruction count reduction from SIMD is the speedup we will see from SIMD, relative to scalar execution.

If SIMD instructions are slower, which is common, then we expect SIMD speedups to be less than the instruction count reduction.

$$SIMD\ efficiency = \frac{\frac{instructions_{scalar}}{instructions_{SIMD}}}{vector\ \ length} \tag{2.1}$$

Example 2.3   Figure 2.3 shows the SIMD efficiency for a program with various fractions of scalar instructions, for various vector lengths. We assume that the vectorized program incurs no additional instructions, scalar or vector, to enable SIMD execution. With this simplification, and under the assumption that SIMD instructions run at the same latency and throughput as their scalar counterparts, this graph is exactly what we get from Amdahl's Law if we apply it to SIMD execution. This simplification and these assumptions are not realiztic, but give us an upper bound on SIMD efficiency, and help illustrate how dramatic an effect scalar code can have. If just 1% of the instructions in an application are scalar, the benefits from SIMD are attenuated significantly for large vector lengths. For very small vector lengths, the impact is modest—with a vector length of two elements, efficiency is 99%. However, the effect grows quickly. With 1% scalar code, for a length of 16 elements, the efficiency drops to 87%, and for a length of 128 elements, the efficiency is under 50%.

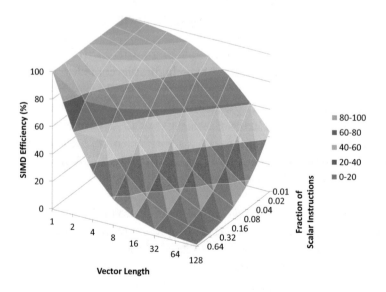

Figure 2.3: Reduction in instruction count from SIMD, and corresponding SIMD efficiency.

Convention 2.4   This book uses several code examples to illustrate scalar and corresponding SIMD computations. The code is in a C-like language, but includes some inline pseudo-assembly to show SIMD instructions. These all begin with the letter "v" to indicate they are "vector" instructions. Register arguments are prefixed with a letter to indicate the type: "v" means vector/SIMD and "m" means mask (described later). Locations in memory are shown as variable names. The destination is always the first argument to an instruction, followed by the inputs. Unless otherwise mentioned, the destination is not an input.

Another important source of inefficiency in SIMD execution is when the number of data items is not an exact multiple of the instruction set's vector length. We typically process a set of data items in a loop, and with SIMD, we can process *VLEN* (i.e., the vector length) items per iteration, as in Figure 2.4. If the number of items is not a multiple of the vector length, we are left with a *remainder* set of items, analogous to the remainder when performing long division.

```
for (i = 0; i < N; i += VLEN)
{
    vload v1, A[i]
    vadd v0, v1, v2
    vstore B[i], v0
}
```

Figure 2.4: Example SIMD loop without handling of the remainder.

Software generally handles the remainder through either scalar execution or predicated execution. For scalar execution, once we detect that we have only a remainder set of items left, we process them one at a time with scalar instructions. For predicated execution, covered in more detail in Section 3.3.2, software can control which results are kept for each instruction. We can predicate each element in the SIMD instructions with a flag indicating whether that element corresponds to a real data item, or one beyond the real set. This allows us to eschew scalar execution completely. However, computing the predicate often requires a comparison instruction, and we need to include that in *all* iterations; thus, even iterations where we handle a full set of data items will include this extra instruction.

Example 2.5   Figure 2.5 shows how loop remainders impact SIMD efficiency. Ignoring looping instructions (e.g., the counter increment and comparison, and the loopback branch), and assuming that the instructions in the loop body are ideally vectorizable, these plots show SIMD efficiency when processing various numbers of data items, for various vector lengths, for both scalar remainder loops and predicated execution.

For the scalar remainder loop, unless the number of data items is an exact multiple of the vector length, efficiency is severely impacted. This effect shrinks as the total number of data items

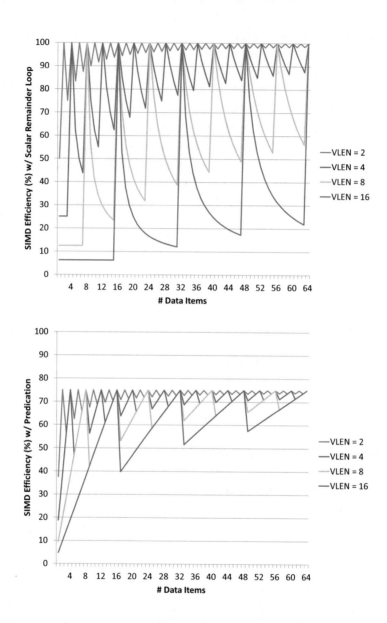

Figure 2.5: Impact of loop remainders on SIMD efficiency, for a scalar remainder loop and for predicated execution.

grows, as the number of SIMD loop iterations increases. With enough SIMD loop iterations, the inefficiency from the scalar remainder loop is eventually amortized.

For predication, maximum efficiency is limited by the overhead of computing the predicate. In our example, we assume a loop with three instructions, and that the SIMD version needs one additional instruction to compute the predicate. Since one fourth of the instructions in the SIMD loop are overhead, the maximum efficiency is 75%, much worse than when using a scalar remainder. However, since we can handle all remainder elements with just one additional loop iteration, rather than multiple scalar iterations, efficiency varies much less than in the scalar remainder loop case. Further, most loops contain more than three computation instructions, better amortizing the predication overhead.

The inefficiency from scalar code outside of a vectorizable loop and from loop remainders are a function of the number of independent data items—we tend to have greater inefficiency with fewer data items. When mapping a parallel computation to modern processors, one more aggravating factor for SIMD efficiency is the mixing of techniques for exploiting parallelism. For instance, in multi-core processors with SIMD support, we typically assign (statically or dynamically) a share of data items to each thread, and each thread can only use SIMD execution on its share of data items. Even if the SIMD efficiency for the full set of data items is quite high, we may force much less efficient SIMD execution by partitioning up the work among threads.

There has been work on flexible SIMD designs to improve SIMD efficiency for codes that map imperfectly to SIMD execution. These designs provide SIMD execution, but allow software to use the execution units independently. Some work dynamically adapts the vector length [Krashinsky et al., 2004, Lee et al., 2011, Park et al., 2012, Rivoire et al., 2006]. The idea is to let software choose how to use the wide execution hardware, e.g., as several threads with narrow execution, or as one thread with wide execution. Other work uses a dataflow design with a grid of execution units that software controls [Govindaraju et al., 2013]; software may group a set of units for SIMD execution. The flexibility these designs provide can improve utilization of wide execution hardware, but this comes at the cost of significantly increased programming and/or compiler effort.

The high sensitivity of SIMD efficiency to scalar code and algorithmic steps with limited data parallelism, combined with greatly improved scalar execution by commodity processors, is what led to a decrease in popularity of vector processors decades ago. However, SIMD is a very effective technique, and the promise of plentiful data parallelism from multimedia applications, scientific computing applications, Big Data applications, and others has motivated a resurgence. SIMD is nearly ubiquitous in general-purpose processors, and SIMD widths have been increasing in recent years. Intel, for example, went from an ISA with 128-bit SIMD (SSE) in 1999, to 256-bit SIMD (AVX) in 2010, and has announced 512-bit SIMD (AVX-512) to first appear in products in 2015.

In high-performance computing, long-vector processors have also made somewhat of a comeback. The Earth Simulator, unveiled in 2002, and the fastest supercomputer in the world until 2004, was a long-vector machine. The follow-on, Earth Simulator 2, was unveiled in 2009. It uses vector processors from NEC that support vectors of 256 elements [JAMSTEC, 2009]. Given the long vector length, to keep operations with little data parallelism from dragging down performance, each CPU also contains a 4-way superscalar execution unit. At its introduction, it was not only the most efficient supercomputer, as measured by LINPACK (a dense linear algebra benchmark), it also delivered the fastest performance in the world on the HPC Challenge Global Fast Fourier Transform (FFT) benchmark [HPCC, 2014].

In addition to scaling SIMD performance by increasing SIMD width, we may increase the number of vector functional units. For designs with short vectors, this has similar benefits, and also presents similar challenges and tradeoffs, as superscalar execution of scalar instructions. For designs with longer vectors, where we may rely on pipelining a single vector through a vector functional unit, we face additional challenges, although solutions like clustering [Kozyrakis and Patterson, 2003] and VLIW Khailany et al. [2008] have been shown to work well.

2.5 PROGRAMMING AND COMPILATION

As SIMD execution is driven by SIMD instructions, a programmer or compiler must place SIMD instructions in a program binary to actually use SIMD hardware. Employing SIMD instructions, either automatically or by a programmer, is a big field of study. Automatic vectorization (*autovectorization*), constructs to help the compiler vectorize programs in existing languages, and languages to expose data parallelism are all areas with extensive amounts of work. Since our focus is on the architecture, we do not attempt to cover these topics in detail. Rather, we give a brief introduction to today's dominant methods for programming processors with SIMD execution.

2.5.1 PROGRAMMING FOR SIMD EXECUTION

A programmer can, of course, write assembly or inline assembly (i.e., inserting an assembly sequence into a program written in a higher-level language) to utilize SIMD execution. Otherwise, the programmer must rely on the compiler to at least some extent to emit high performance SIMD code.

In addition to the normal duties of a compiler for a scalar architecture, a compiler for a SIMD architecture must analyze the code for parallelism, similar to an automatically parallelizing compiler [Hiranandani et al., 1992, Padua et al., 1993]. This allows it to determine which proximate operations and/or operations inside a loop match the SIMD instructions provided by the target instruction set and can execute safely in a SIMD manner.

The compiler's parallelism analysis depends on the assumptions and restrictions of the language in which a program is written. While languages that encourage or even force the programmer to explicitly expose parallelism are increasingly popular, especially for graphics processing units (GPUs), today's most common choices for programming CPUs with SIMD execution are scalar languages.

Some scalar languages are easier than others to analyze for vectorization. Fortran, in particular, has a long history of being used for applications mapped to SIMD hardware, and a similarly long history of vectorizing compilers, starting with work from David Kuck [Kuck et al., 1972] and Cray Research [Cray, 1977]. On the other hand, programs written in the extremely popular C and C++ rely heavily on the compiler to analyze the parallelism in the code; these languages are very flexible, which may be convenient for programmers, but makes it difficult for compilers to analyze. We briefly describe the main challenges in static analysis for vectorization in Section 2.5.2.

Since C and C++ are so popular and yet so challenging, several compiler vendors have introduced extensions for C and C++ (and other languages) to provide programmers with a means to convey dependence information, and other helpful hints, to the compiler. These extensions cover a spectrum of levels of abstraction.

Instrinsics are the extension method closest to writing assembly. A compiler intrinsic is a function call that maps to a specific instruction or a small sequence of instructions. These may require adding data types to the language for parameters and/or the return value, such as types that map to SIMD registers. While a programmer essentially directs the compiler to use certain instructions with intrinsics, he/she still relies on the compiler for optimizations such as register allocation and instruction scheduling. Intel's icc supports intrinsics for SIMD instructions and other instructions a programmer may want to help the compiler use [Intel, 2014a]. Microsoft's Visual Studio supports intrinsics for x86 and ARM [Microsoft, 2014]. Finally, gcc supports intrinsics (called "built-in functions") for many architectures, including SIMD support for ARM, X86, PowerPC, and SPARC [FSF, 2014].

A similar approach can be provided by C++ SIMD classes that include data types and operators for data parallel operations [Intel, 2014b]. This allows programmers to write code that appears scalar (e.g., x + y) but in fact directs the compiler to perform a SIMD computation. For even more flexibility, Cilk Plus elemental functions [Intel, 2014b] allow a user to specify his/her own data parallel operations.

Finally, pragmas are a portable approach that are more decoupled from the source code [Intel, 2014b]. A programmer may specify certain properties of a loop or other logical block of code by placing a pragma statement just before it. Pragmas are essentially hints, as opposed to intrinsics or C++ classes, which *must* be recognized and honored in some way by the compiler to produce a binary that will produce a correct answer. As hints, they can be completely ignored, and the remaining code is a correct scalar program.

Pragmas may take several forms. First, they may directly assure a compiler that certain dependences (e.g., cross-iteration dependences) don't exist, helping or even overriding the results of alias analysis. Second, they may inform a compiler that an operation in the block fits a known dependence pattern (e.g., a reduction operation—see Chapter 5). Finally, they may influence one of the compiler's many cost functions for optimization, such as suggesting the number of iterations of a loop to optimize for, when this cannot be determined statically.

While some compilers extend popular scalar languages to allow programmers to assist in vectorization, some languages require that programmers express parallelism in certain forms that are easy for compilers to map to parallel hardware, including SIMD hardware. OpenCL is a language based on an open standard designed explicitly for parallel programming of a variety of processors, including both CPUs and GPUs [Khronos, 2013]. Coarse-grained data parallelism is mapped to a grid of work items, and fine-grained parallelism is mapped to built-in vector data types and functions, analogous to intrinsics or C++ SIMD classes. Nvidia's CUDA popularized this approach of explicitly parallel programming, and is very similar to OpenCL in many respects [Nvidia, 2014]. Other languages with inherent parallelism support spring from the HPC community, such as Chapel [Chamberlain, 2013], Fortress [Oracle, 2011], and X10 [Saraswat et al., 2014].

2.5.2 CHALLENGES OF STATIC ANALYSIS

An autovectorizing compiler faces many challenges in determining the dependences and control flow of a program at compile-time with sufficient precision to enable the *beneficial* use of SIMD instructions. We briefly describe these challenges below. These are fundamentally SIMD architecture issues rather than compiler issues—with sufficiently aggressive hardware that performs dynamic dependence and control flow checks, we could enable correct execution of a vectorized program that completely ignores any or all of these issues as long as the expected behavior was unambiguous. However, at least today, reasonable cost hardware requires that the compiler be conservative.

For a naively written program, where the programmer has made no attempt to write his/her code with an eye for vectorization, the compiler often has to be so conservative that it can vectorize very little code. This has more to do with the restrictions, or lack thereof, in the programming language than with the level of sophistication of the static analysis. It also has to do with data layout and other decisions that a programmer makes that are difficult for a compiler to override. With a program written to be more amenable to vectorization, modern autovectorizing compilers can come close to the best that a human expert can do [Satish et al., 2012].

Determining where/what to vectorize

For a given computation, which data items should be grouped into vectors? For example, at which level of a nested loop should vectorization be attempted, or should it be attempted even within a single loop body (e.g., for manually unrolled loops)? In addition to understanding the parallelism in the code, the compiler typically wants to know the trip count (i.e., number of iterations) of each loop, to understand maximum vector length and how well any overheads will be amortized. Initially, compilers only considered vectorizing innermost loops to keep these analyses relatively simple, but that barrier was broken and more recent work has enabled vectorization in even more cases [Nuzman and Zaks, 2008, Trifunovic et al., 2009].

Determining if all data items have the same control flow

Since SIMD execution requires that all data items in a vector follow the same control flow, the compiler must determine if this is the case in the original scalar code. It may not be for a variety of reasons, such as if-checks with input-dependent conditions, or vectorization of an outer loop with variable trip count for the inner loop. In some cases, loop transformations may eliminate control flow differences [Hanxleden and Kennedy, 1992]. To correctly handle divergent control flow, the compiler may test all elements in a vector at once and jump to scalar code if any elements follow a different path from the others [Shin, 2007]. The compiler may instead use some conditional SIMD operations to convert control flow into data flow, enabling SIMD execution at the cost of some efficiency [Bik et al., 2002]. It's also possible for control flow to be consistent across data elements, but for there to be challenges in determining this, such as function calls; sophisticated analysis may be required in these situations [Tian et al., 2012].

Determining if data is aligned

Some SIMD architectures restrict memory operations to addresses aligned according to the SIMD width. Originally, vectorizing compilers for such architectures needed to determine that all data was aligned before vectorizing a loop. This constraint has been made less onerous with sophisticated static alignment analysis, and with techniques to dynamically align data structures, albeit with some overhead [Bik et al., 2002, Eichenberger et al., 2004, Wu et al., 2005].

Understanding how data is organized

Code operating on data structures that inherently require non-contiguous accesses, or algorithms that rely on strided or more complex access patterns such as the butterfly pattern in FFT, may be vectorized, but require that software carefully move data into and out of vector registers. Significant work has gone into understanding data access patterns to successfully vectorize such computation, and to minimize the overhead of data movement [Kim and Han, 2012, Nuzman et al., 2006, Ren et al., 2006].

Determining the safety of vectorizing the memory operations

The compiler may violate correctness by vectorizing memory accesses if the resulting code either accesses a memory location that the scalar code would not have, or violates a data dependence. For loops with control flow, the compiler must ensure that no memory locations are touched unsafely. For example, conditional reads to elements in an array may seem safe to perform unconditionally in a vectorized manner, but this may trigger a fault that the original code would not have because the programmer carefully chose a condition to guard against that possibility. Vectorizing memory operations also can change the order of the original scalar memory operations, which may violate data dependences. Static data dependence analysis via Banerjee's test or another test can sometimes determine if reordering memory operations via vectorization is safe [Banerjee,

1976]. Dynamic testing for overlap of memory references can enable vectorization of even more loops [Bik et al., 2002, Bulic and Gustin, 2004].

CHAPTER 3

Computation and Control Flow

In this chapter, we describe the in-core, non-memory components of SIMD execution: data storage (i.e., registers), computation instructions, and control flow.

3.1 SIMD REGISTERS

SIMD instructions operate on data items either in memory or in registers. We cover memory in Chapter 4. For data in registers, the data items might be kept in dedicated SIMD registers (e.g., in Intel's SSE and AVX) or across multiple general-purpose registers (e.g., Sun's VIS). A dedicated SIMD register file requires significantly more area, but allows for a clustered design of the core—the SIMD functional units and registers can both be separated from the scalar portion of the core.

Figure 3.1 shows a scalar and SIMD register file. The scalar registers are physical registers, so we may have many more than we have architectural registers. Each SIMD register is wider than a scalar register, which is typically 32 or 64 bits wide (64 in our figure). This allows a SIMD register to hold multiple scalar data elements. Due to their size, we may limit the number of

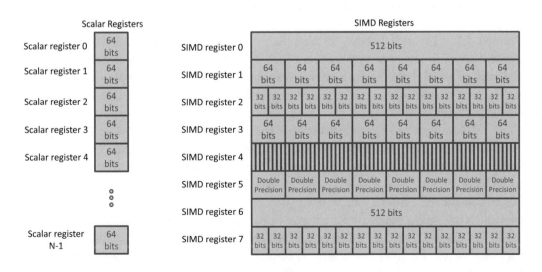

Figure 3.1: Scalar and SIMD register files. The SIMD registers hold a variety of data types: long bit vectors; 64-bit, 32-bit, and 8-bit integers; and double-precision floating-point numbers.

SIMD physical registers, or eschew register renaming altogether. In our example, SIMD registers are 512 bits wide; one may hold 64 values that are each an unsigned byte, while another may hold 16 single-precision floating values. The data type that a SIMD register holds is restricted only to what SIMD instructions expect as inputs or outputs.

Software generally tracks the contents of a SIMD register, treating them as being of a certain data type, but hardware does not track this. As far as hardware is concerned, the data type in a SIMD register is a temporary designation, and is determined by the instruction that uses it as an input or output. For example, a logical instruction such as an AND may produce some output register that software interprets as 32-bit integers, but those bits are just a long bit vector, as far as the hardware is concerned. A later instruction may use the same register as an input to an addition instruction that interprets the contents as 32-bit integers, or some other data type.

Similarly, a SIMD register can hold a mix of data types, rather than holding a homogeneous set of data elements. However, software rarely desires this since SIMD instructions perform a single operation on a set of input data items; thus, instructions that interpret a SIMD register as a mix of data types are rare.

3.2   SIMD COMPUTATION

3.2.1   BASIC ARITHMETIC AND LOGIC

The fundamental building block of SIMD instruction sets is a group of basic instructions that operate on multiple data items. These include arithmetic operations (addition, subtraction, multiplication, and division), boolean operators (e.g., AND, OR, XOR), logical and arithmetic shifts, data type conversion, and data selection operations (e.g., min and max).

Example 3.1   Figure 3.2 shows an example instruction sequence comprising basic arithmetic instructions. This sequence uses six input SIMD registers, some of which hold 64-bit integers, and some of which hold double precision floating point values. The initial values in the registers are shown at the top of the code sequence. It produces a single SIMD register output, containing double precision values.

The first operand to each instruction is the destination, so in this example, all five instructions place their results in register v0. The first instruction in the sequence, vpaddq, performs a 64-bit integer addition on registers v0 and v1. The second instruction, vpxorq, performs a logical XOR on v0 and v2. As XOR is a logical operation, the data type does not matter; however, the instruction still carries a data type for predication purposes. The third instruction, vpmullq, performs a multiplication of 64-bit integers in v0 and v3, and keeps only the low 64 bits of each result. The fourth instruction, vcvtqq2pd, converts 64-bit integers in v0 to double precision floating point values. The final instruction, vfmaddpd213, performs a fused multiply-add of double precision floating point values—it first multiplies values in v0 and v4 and then adds the result to values in v5. As this instruction takes three inputs, the destination register doubles as an input register.

```
// Initial values
// v0 holds 8 64-bit integers: {     5,    24,     3,     1,    17,     0,     8,     2}
// v1 holds 8 64-bit integers: {     4,    11,    20,     7,     9,     1,    26,     2}
// v2 holds 8 64-bit integers: {    10,     2,     5,    20,     3,     4,     0,     4}
// v3 holds 8 64-bit integers: {    20,     1,     2,     0,     2,    11,     1,    36}
// v4 holds 8 doubles:         {   0.1,   2.0,   0.5,  88.7,   0.9,   0.1,   1.6,  26.2}
// v5 holds 8 doubles:         {   7.3,   3.4,  10.0,   0.3,   2.1,   1.2,  17.3,  32.6}
vpaddq v0, v0, v1        // v0 = {     9,    35,    23,     8,    26,     1,    34,     4}
vpxorq v0, v0, v2        // v0 = {     3,    33,    18,    28,    25,     5,    34,     0}
vpmullq v0, v0, v3       // v0 = {    60,    33,    36,     0,    50,    55,    34,     0}
vcvtqq2pd v0, v0         // v0 = {  60.0,  33.0,  36.0,   0.0,  50.0,  55.0,  34.0,   0.0}
vfmaddpd213 v0, v4, v5   // v0 = {   7.9,  69.4,  28.0,   0.3,  47.1,   6.7,  37.7,  32.6}
```

Figure 3.2: Example sequence of basic SIMD arithmetic operations. The instruction mnemonics (but not the register names) are from Intel AVX-512, which include an indication of the data types to assume that the SIMD registers hold. A "p" after the "v" indicates an integer instruction, and later letters are more specific: "q" means 64-bit integers and "pd" means double-precision floating point values.

Most implementations of SIMD instruction sets include ALUs wide enough to operate on all data items in a SIMD register simultaneously. Thus, if the latency of all of the instructions in the example was a single cycle, the entire sequence could execute in five cycles. However, some implementations use narrower ALUs, and need to pipeline execution of each instruction, akin to a vector processor.

Intel's Pentium III is a concrete example of this. It included the first implementation of Intel SSE, 128-bit SIMD that comprised support only for single-precision floating point. The ALUs in the Pentium III were 64 bits wide, however. Thus, each SSE instruction needed to occupy an ALU for two cycles to execute a single SSE instruction. Unlike processor designs where this pipelining is done to allow high frequency operation, the Pentium III (and some other, more modern designs) used pipelining to allow a cheaper ALU implementation.

3.2.2 DATA ELEMENT SIZE AND OVERFLOW

Since SIMD instructions operate on a fixed number of bits, the number of data elements a single SIMD instruction operates on is inversely proportional to the size of each element (as illustrated in Figure 3.1). Therefore, to maximize performance, programmers should use the smallest element size possible. For example, 16 bits, or even 8 bits, is often enough to hold values for many multimedia applications.

Forcing values to fit in compact storage can have a downside, though—overflow. Some operations, like addition and multiplication, naturally have larger results than operands. For addition, the result is only a single bit larger, but for multiplication the result is twice as large as the operands. The vast majority of SIMD arithmetic instructions use the same data type for the output as for the input(s). Thus, if the result of the arithmetic operation cannot be represented in the number of bits the programmer has allotted for an operand, we will have overflow, and risk losing information.

Instruction sets commonly include support for scalar overflow; e.g., for addition, many ISAs include a carry bit, and for multiplication, some ISAs, such as x86, either place the result in a register that is twice as large as the operands, or place the low half of the result in one register and the upper half in another register. Instead of providing a SIMD equivalent of the carry bit, one common feature is to provide *saturating arithmetic* instructions. These instructions, in the presence of overflow, will clip the result value to the maximum or minimum representable value. For multiplication, a common SIMD approach is to split multiplication into two instructions: one to obtain the low half of the result, and one to obtain the high half of the result. The reason for this approach is that in many cases, an application only needs the low half of the multiplication result. However, in rare cases where it needs the full result, this approach forces software to use two multiplication instructions plus additional instructions to combine the two halves.

3.2.3 ADVANCED ARITHMETIC

While basic arithmetic instructions suffice to exploit data parallelism in many applications, some application developers desire higher performance for some common, more complex operations. Some SIMD instruction sets have been enhanced with support for these operations.

One class of these operations is transcendental functions and other nonlinear functions. Recently, SIMD development has been driven largely by scientific applications, and other high-performance computing applications. Many of these applications use more complex mathematical operations than addition and multiplication. Reciprocal, square root, reciprocal square root, sine, cosine, and tangent are often needed, but are slow to compute using a sequence of traditional arithmetic instructions—these are normally computed using lookup tables and/or evaluating several iterations or terms in a numerical approximation method, such as Newton-Raphson [Cornea-Hasegan et al., 1999]. Instructions explicitly for these operations can be faster than a software sequence, but typically use the same underlying implementation: lookup tables and/or a multi-operation implementation. This makes them relatively expensive and/or slow.

To address the performance and area limitations of instructions for computing complex arithmetic functions, some instructions sets include *approximate* arithmetic operations. These perform the same mathematical operations as the transcendental and other nonlinear functions, but at reduced precision. To get an acceptable final result, not all steps in an algorithm using floating point computations necessarily need to produce answers at the full precision specified by the data type. For example, a large computation may use double precision floating point values throughout, but a part of the computation may produce just a single bit answer, such as whether some condition has been satisfied. In cases like this, approximate mathematical functions are a good fit—they may provide *enough* precision and significantly improved performance. Reciprocal square root is an example arithmetic operation that is present in many SIMD instruction sets in an approximate form (e.g., ARM's NEON, IBM and Motorola's AltiVec, and Intel's SSE/AVX/AVX-512). Implementations of approximate arithmetic instructions tend to use the same approach as for exact arithmetic instructions, but with either reduced cost (e.g., smaller lookup tables) or fewer steps or terms. There has been considerable recent interest in approximate computing beyond SIMD, including in the architecture community. For example, Venkataramani et al. [2013] proposes instruction set support for operations with reduced precision, which are similar in many respects to SIMD reduced precision instructions, but more extensively applied.

Another class of advanced arithmetic instruction is one that operates *horizontally*. Horizontal operations combine multiple data items from the same vector. For example, if we have a vector comprising a set of numbers, and we want the sum of those numbers, then we may use a *horizontal add* operation that sums all of the elements in a vector. Horizontal instructions are fundamentally different from other SIMD instructions in that they introduce data dependences between different elements in a vector. This violates the SIMD premise of operating on independent data items. It can also tie a piece of software to a specific SIMD width, beyond finding the right number of SIMD iterations. However, these operations are needed so frequently in real algorithms, that they are often included in some form in SIMD instruction sets.

Horizontal arithmetic operations are not scalable. For example, adding all of the elements in a vector together, even in an efficient binary tree fashion, requires $\log_2 VLEN$ additions. Thus, SIMD instruction sets tend to include few horizontal operations, and most of those are generally data movement operations, to move or copy an element from one position in a vector to another; data movement between different vector positions allows arbitrary horizontal arithmetic operations without forcing software to write data to memory and then read it back. Horizontal operations are a large topic, which we cover in depth in Chapter 5.

3.3 CONTROL FLOW

Control flow in programs with SIMD instructions is the same as in purely scalar programs—loops, function calls, and other scalar control flow are fundamental operations, even in algorithms with data parallelism. However, control flow presents a problem when we want to execute completely different code paths for different data elements, or even just certain SIMD instructions for only a subset of data elements. The general problem of wanting to perform different operations on different elements in a vector is known as *control divergence*.

3.3.1 SIMD EXECUTION WITH CONTROL FLOW

Control divergence springs from SIMD execution's requirement that we perform exactly the same sequence of operations on all data items. Any deviation from that restriction introduces control divergence. For programs in languages like C, control divergence most often springs from conditionals that may evaluate differently for different items in a vector. These may take the form of if-checks, inner loops within a vectorized outer loop, or more complex constructs like switch statements.

Example 3.2   Figure 3.3 shows an example vectorizable loop containing an if-check that may go different directions in different iterations.

```
// Scalar loop
for (i = 0; i < N; i++)
{
    if (A[i] > 0)
    {
        B[i] += C[i];
    }
}

// Simple vectorized version
for (i = 0; i < N; i += VLEN)
{
    vload v0, B[i]
    vload v1, C[i]
    vadd v2, v0, v1
    vstore buffer, v0

    for (j = 0; j < VLEN; j++)
    {
        if (A[i + j] > 0)
        {
            B[i + j] = buffer[j];
        }
    }
}
```

Figure 3.3: Example vectorizable loop with an if-check that may lead to control divergence.

To use SIMD execution in such situations, we need independent control of the operations performed on the data items in a vector. One method that requires no additional hardware support is to use SIMD execution to speculatively execute the computation inside the if-check for all elements in a vector, write the output of the computation to a temporary buffer, and use scalar execution to conditionally "commit" the output (i.e., conditionally write each updated element of B to memory). The second loop in the example shows this technique applied to the first loop.

Note 3.3   Depending on the semantics of the original programming language, this transformation may not be safe/legal, since it may cause us to access memory locations that the original scalar execution would not. For example, we may read some B[i] and C[i] for which A[i] is zero. This could trigger errant faults. Therefore, for this and other methods of handling control divergence, the programmer may need to perform the transformation themselves, or give the compiler permission to perform the transformation (e.g., via a pragma).

The above technique introduces two types of scalar operations for the commit phase: (1) evaluating the condition and (2) copying the output from the buffer to its destination. We can accelerate the first type of scalar operation with SIMD comparison instructions. These perform a comparison on all data items in a vector simultaneously. Figure 3.4 shows an example SIMD comparison instruction that compares each element of input A with the corresponding element of input B, testing for $A > B$. Unlike scalar comparisons, it is not obvious how we should handle the output for SIMD comparison instructions.

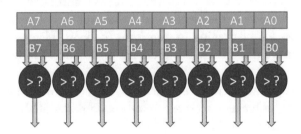

Figure 3.4: SIMD greater-than comparison.

In some instruction sets, scalar arithmetic and/or compare instructions set flags to indicate certain properties of the result, such as a zero result. Conditional branch instructions then use one or more of these flags to determine whether to branch.

In contrast, SIMD comparison instructions perform a specific comparison between pairs of data elements, and then output the single bit result of that comparison (i.e., true or false). The output may go into a register dedicated to holding these results, e.g., as a bit vector, or it may go into a SIMD register. For example, Intel SSE comparison instructions place comparison results into SIMD registers in an encoded way: if the comparison is true for a pair of elements, the corresponding output element has all of its bits set to one, otherwise, all of the output element's bits are cleared to zero.

SIMD comparison instructions by themselves will not eliminate the if-check in the commit phase since we still conditionally copy results. We must execute each scalar copy conditionally, by moving each comparison result into a scalar register and either using that as the control for a conditional move, or testing it for pass/fail (e.g., with a scalar if-check). However, SIMD comparison instructions allow us to perform and efficiently combine the results of multiple comparisons.

Example 3.4   For example, Figure 3.5 shows an example vectorizable loop containing a nested if-check. The second loop is vectorized using SIMD comparison instructions. We test the two conditions, $A > 0$ and $B < 0$, with one SIMD comparison instruction each, and then we combine the tests with a logical AND. The scalar conditional copy loop still has an if-check, but only one, and it is a simple pass/fail check.

The conditional copy remains a significant source of inefficiency: it is scalar code, it requires us to move SIMD computation results and comparison results into scalar registers, and it forces us to re-test the condition for each element. In the next section, we discuss conditional SIMD execution, which addresses this. Before we solve this problem, we describe additional sources of control divergence.

For nested loops, we may elect to vectorize the outer loop rather than an innermost loop, e.g., if the inner loop has a small number of iterations or is not data parallel. However, when we vectorize an outer loop, there is a possibility that the trip count for the inner loop will be different for different elements of a vector; the result is control divergence. In theory, this can be handled with an extension to the conditional scalar copy technique described above. In some cases, software can use loop flattening to transform the loop [Hanxleden and Kennedy, 1992] and eliminate the control divergence, but there are restrictions. Other compiler work has also addressed this problem [Nuzman and Zaks, 2008, Trifunovic et al., 2009], but still carries restrictions. When these analyses and transformations fail, unless we can prove that the inner loop trip count is consistent, or the architecture supports conditional SIMD execution, it is generally better to avoid outer loop vectorization.

Example 3.5   Figure 3.6 shows a nested loop with data parallelism across iterations of the *outer* loop. In this example, the values in A are the trip counts for the innermost loop. If the values in A are not all the same, and we vectorize the outer loop, we may have control divergence. For instance, suppose A[0] is 3 and A[1] is 4. In our first iteration of the vectorized outer loop, we will, in SIMD fashion, work on elements 0 and 1 (and possibly others) of A, B, and C. To do so correctly, we must either suppress execution of a fourth iteration of the inner loop for element 0 of the vector, or stop SIMD execution after the third iteration of the vectorized inner loop, and handle the fourth iteration for element 1 of the vector in scalar fashion.

A final source of control divergence, which we discussed earlier (in Section 2.4), is SIMD instruction sets having a fixed vector length. Original vector processors like the Cray-1 used vector length registers to indicate the number of valid elements in a vector [Cray, 1976], and so could perfectly match vector length to the number of data items. However, modern, short vector SIMD instruction sets have a fixed vector length. If the number of data items for a computation is not a multiple of the vector length, then we may use straightforward SIMD execution for all but the

```
// Scalar loop
for (i = 0; i < N; i++)
{
    if (A[i] > 0)
    {
        if (B[i] < 0)
        {
            B[i] += C[i];
        }
    }
}

// Vectorized version with SIMD comparisons
for (i = 0; i < N; i += VLEN)
{
    vload v0, A[i]
    vxor v1, v1, v1 // set v1 to 0
    vcmpgt v2, v0, v1

    vload v3, B[i]
    vcmplt v4, v3, v1
    vand v5, v4, v2
    vstore cond, v5

    vload v6, C[i]
    vadd v7, v3, v6
    vstore buffer, v7

    for (j = 0; j < VLEN; j++)
    {
        if (cond[j])
        {
            B[i + j] = buffer[j];
        }
    }
}
```

Figure 3.5: Example vectorizable loop with a nested if-check that may lead to control divergence.

```
for (i = 0; i < N; i ++)
{
    for (j = 0; j < A[i]; j++)
    {
        B[i] += C[i];
    }
}
```

Figure 3.6: Example vectorizable loop with a nested loop that may lead to control divergence.

last few data items; these are the *remainder*. We have two choices for dealing with the remainder. First, we can add a scalar remainder loop. Second, we could modify the primary SIMD loop to handle all data items in SIMD fashion, vectorizing the remainder. Here, we add a conditional inside the loop to test whether the iteration number for that data item is beyond the loop bound (i.e., number of valid items). A vectorized remainder is generally faster if we have support for conditional SIMD execution.

3.3.2 CONDITIONAL SIMD EXECUTION

Being forced back into scalar execution when we have control divergence is both inefficient and inconvenient for software. We can avoid this with conditional SIMD data movement, also known as a blend operation, or merge instruction, e.g., on the Cray-1 [Cray, 1976]. A blend is implemented with a multiplexer—it takes two data inputs for each vector element, and a control input that selects, for each vector element, which input element to pass as the output. For some blend instructions, the control input is restricted to an immediate value; such a restriction severely limits its usefulness for control divergence since this requires compile-time knowledge of which data item we want for each element in the vector, or for us to construct a set of possible combinations and branches to lead us to the right one. A blend that takes dynamic control information is far more useful. Since the control input will most often come from a SIMD comparison instruction, the form it takes (e.g., a regular SIMD register) should match the output of a SIMD comparison instruction. Figure 3.7 shows the operation of a blend instruction with data inputs A and B, and control input C.

Blend instructions allow us to use SIMD execution for computations with potential control divergence, but they still require us to use explicit instructions for conditional data movement. For computations with few instructions on the divergent path, and/or many possible paths that all need to be merged together, this may result in significant instruction overhead.

Masking

Some architectures provide for implicit blending via the use of *masking*, a SIMD form of predication. A predicated scalar instruction takes as an input a single bit indicating if the instruction

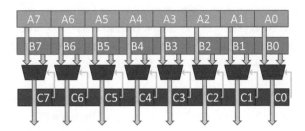

Figure 3.7: SIMD blend instruction.

should actually be executed. Similarly, in SIMD predication/masking, a SIMD instruction takes a set of bits, one for each vector element, indicating if the instruction should be executed for that element in the vector. Predication was developed to convert control dependences into data dependences [Allen et al., 1983], and was implemented as vector masking in several early vector processors such as the Hitachi S-820 [Eoyang et al., 1988], the Fujitsu VP-100/VP-200 [Miura and Uchida, 1983], and the NEC SX-2 [Watanabe, 1987].

Masking, when supported, can be applied to most SIMD instructions. For example, we may provide a mask for a SIMD addition instruction if we want to perform the addition conditionally. Also, we may use masking for a SIMD data movement instruction to get the equivalent of a blend instruction.

Masks are bit vectors, and may come either as outputs of SIMD comparison instructions, or as pre-computed values. As we discussed for SIMD comparison instructions, an architecture may store such bit vectors in regular SIMD registers, encoded in some way, in scalar registers, or in dedicated *mask registers* [Cray, 1976]. Dedicated mask registers carry some advantages. First, we have more registers left for scalar or SIMD data. Second, compared to using SIMD registers, we waste less space storing masks in dedicated registers—mask registers need only be as many bits large as the largest possible vector length. Third, we can avoid adding another read port to the scalar or SIMD register file; since the mask is another operand, we need to read it from the register file before execution. Highly ported structures are expensive, and we already need at least two read ports (for data operands) to the SIMD register file.

If masks are held in a separate register file, and these registers are architecturally visible, then we may want instructions to directly manipulate them. Since there are a finite number of such registers, it is useful to be able to spill and fill them to or from memory. It may also be useful to combine the results of conditionals via logical operations on masks, such as ANDing or ORing them together. All of these options could be provided by moving masks to/from scalar registers, performing the operation there, and then moving the results back; however, this carries overhead of moving things back and forth, and also uses up scalar registers, which partly defeats the purpose of having separate mask registers.

Masking need not be explicit. In particular, some GPUs use *implicit masking*, where mask registers are not architectural [Fung et al., 2007]. Implicit masking is not GPU-specific, but rather requires a parallelization model where control flow is restricted such that the hardware can automatically compute and use masks. The benefit is that we do not need extra instructions to set, manipulate, save, or restore masks.

It is fairly obvious how a mask bit can/should be interpreted for conditional data movement instructions, but it is less obvious for other instructions. Perhaps the most intuitive meaning of a mask bit that is zero is to suppress the instruction for that vector element, and leave the contents of the destination SIMD register alone for that vector element. This is called *blending masking*, since it mimics a blend operation. However, in many cases, software uses a mask bit of zero for elements that it doesn't care about at all, and we can change the meaning to get better performance on some microarchitectures.

Example 3.6   Figure 3.8 shows a scalar loop and its vector equivalent, where we use masking for both adds in the loop, as well as the loads and store to memory. Masking on the additions has no impact on the outcome of this loop.

When software uses masking for elements that it doesn't really care about, blending masking can create an unwanted input dependence. Depending on the execution pipeline and relative execution times and latencies of the instruction involved, hardware may not know the value of the mask ahead of time. To be safe, it must assume that some of the destination register's values will not be overwritten; this means the destination register is also an input register. For cores with relatively low instruction throughput, such as narrow-issue cores using in-order execution, this is not a problem. However, for higher throughput cores, such as wide-issue superscalar cores using out-of-order execution, this can severely limit performance—tight SIMD loops with independent iterations can gain cross-iteration dependences.

A solution to this is another form of masking, where we overwrite all elements of the destination register, ensuring that it is not needed as an input for blending. The most obvious choice is to zero out elements with a corresponding mask bit of zero—this is known as *zeroing masking* [Intel, 2014c]. Instruction sets with only zeroing masking or only blending masking face some drawbacks: needing to sometimes use explicit blend instructions or sometimes incurring unwanted input dependences. A solution to this is to provide both, letting software choose which form of masking it wants.

We implied earlier that using blending masking when it doesn't affect the final result carries negative consequences, since it can cause performance issues. That, however, ignores energy consumption. If we suppress the computation for a masked-out element [Keryell and Paris, 1993], then we may save significant energy when using masking. This requires that we know the value of the mask when we do the computation. We may not have it, depending on the pipeline and precise timing of the execution of the instructions involved. In that case, to suppress execution of masked-out operations, we need to wait for the mask input before performing a computation. Waiting for the mask has significant disadvantages: if this instruction is on the program's critical

```
// Scalar loop
for (i = 0; i < N; i++)
{
    if (A[i] > 0)
    {
        B[i] += A[i] + C[i];
    }
}


// Vectorized version with masking
for (i = 0; i < N; i += VLEN)
{
    vload v0, A[i]
    vxor v1, v1, v1
    vcmpgt m0, v0, v1

    vload v2, m0, C[i]
    vadd v3, m0, v0, v2

    vload v4, m0, B[i]
    vadd v5, m0, v3, v4
    vstore B[i], m0, v7
}
```

Figure 3.8: Example vectorizable loop, shown vectorized with masking. Mask registers are denoted with an "m."

path, then this may reduce performance, which may also *increase* energy consumption. Alternatives are to opportunistically use the mask if it happens to be ready when the other operands become available, or to apply masking at the *end* of an instruction, when possible. To apply a mask at the end of the instruction, we execute the operation specified by the instruction as soon as the data inputs are ready, and then use a final logic stage to conditionally write the speculatively computed result to the destination register.

Masking can save us energy by skipping unnecessary operations, but it also carries hardware and energy overheads. If we have a dedicated mask register file, it requires area in the core. We also need area for the logic for mask manipulation instructions. Masked instructions need to include the mask operand, which makes for larger instructions, and for more complex instruction decoders. Finally, we spend energy on every masked instruction to read the mask register. In most programs, most SIMD instructions do not need to be masked; therefore, to avoid having to produce a mask with all elements enabled, and also to reduce the energy overhead of reading the mask, some instruction sets include either unmasked versions of SIMD instructions or a special mask register specifier that indicates "all ones."

Conditional Memory Operations

Conditionally reading from or writing to memory is a little different from conditional operations on registers. As explained earlier, one implementation of masked operations is to perform the operation and *then* apply masking. However, for memory operations, that can carry some side effects. In particular, unless we want to complicate the programming model significantly, we cannot speculatively access memory locations that may trip faults.

Masked memory instructions, or blend instructions that read or write memory, may provide *fault suppression* semantics [Intel, 2014c]. Conditional reads from memory may access all locations specified by the instruction, but have the faults suppressed if they are caused by accesses to masked-out elements. For SIMD memory accesses to contiguous addresses, this typically happens at the beginning or end of a vector, when we may be masking out elements before the beginning or after the end of an array. Assuming a memory system relying on paging for permissions, for most types of faults, we are guaranteed that they will only occur on an access that crosses a page boundary, and we could use simple fault suppression hardware that only handles this case. For accesses to non-contiguous addresses, however, we may need fault suppression on any element in the vector.

Fault suppression is a powerful technique that allows a safe single SIMD code path for all elements. Without it, we sometimes need branches to check the legality of addresses before at least the first SIMD memory instruction in each basic block. Fault suppression can also be useful for arithmetic operations that can trip faults, such as division (e.g., division by zero).

3.3.3   EFFICIENCY IMPLICATIONS OF CONTROL DIVERGENCE

In the previous section, we described instructions and hardware support for conditional SIMD execution. With those capabilities, we can often turn a SIMD computation into a single basic block, regardless of the complexity of the original control flow. However, while the SIMD code path may be simple, that doesn't mean performance is high.

Control divergence reduces SIMD efficiency, by forcing us to waste vector elements in SIMD instructions. For example, if only half of the elements in a vector pass a condition, then the conditional SIMD instructions that follow have, at best, a 50% efficiency—half of the elements in the vector are essentially in a "do not care" state, since we will discard their values. The fundamental performance effect of control divergence is that as the amount of control divergence rises, i.e., as the fraction of elements that passes a conditional check drops, the SIMD efficiency of the following code drops. In programs with nested conditionals, unless all vector elements take the same path for some conditionals, with each level of nesting, we see reduced SIMD efficiency.

Figure 3.9 illustrates this effect. It shows the SIMD efficiency assuming infinite SIMD width, for a theoretical computation, for various fractions of elements that pass a conditional

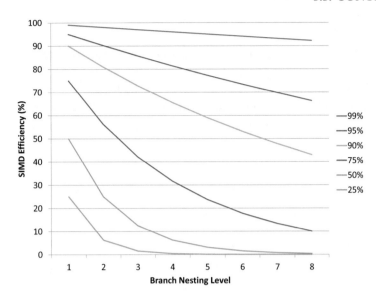

Figure 3.9: SIMD efficiency vs. nesting levels of branches, for various percentages of elements that pass each branch condition.

check, for various levels of branch nesting. For conditional checks that incur very little control divergence (i.e., 99% of the elements pass), we maintain good SIMD efficiency, even for several levels of nesting. However, even a small amount of control divergence can quickly be magnified by branch nesting—a 95% pass rate for each condition results in a SIMD efficiency of only 81% at four levels of nesting. If individual branches introduce a large amount of control divergence, then we may see better performance from scalar execution, especially if the branches are nested. For instance, with a 25% pass rate per condition, SIMD efficiency is at only 6% with two nested conditions.

In practice, control divergence is not quite so catastrophic to SIMD efficiency. One big mitigating factor is that SIMD efficiency is limited by vector length. In the presence of possible control divergence, we can always check to ensure we have at least one valid element before executing a SIMD computation. If no elements are valid, we can branch around the computation— we actually have no control divergence in this case, since all elements want to follow the same path [Shin et al., 2004]. Thus, the lower bound for SIMD efficiency is $\frac{1}{VLEN}$, ignoring the overhead of the additional check and branch. For instance, with a vector length of four elements, the lower bound for SIMD efficiency is 25%.

This is the main reason why, for SIMD instruction sets supporting only short vectors, designers have generally ignored control divergence. Intel, for example, waited until SSE4 to introduce blend instructions, and did not introduce masking until AVX-512, once vectors were 512 bits long.

Another reason control divergence does not doom us to low SIMD efficiency is that the elements that pass a condition are not necessarily uniformly distributed. This may be a natural property of the computation or input; e.g., the first several elements may be much more likely to pass if we are checking for a boundary condition. Software may also intentionally bias the distribution to reduce control divergence in each vector; we discuss this further in Section 5.4.

CHAPTER 4

Memory Operations

SIMD compute instructions operate on multiple data items at a time; to feed these, we need memory instructions that similarly read or write multiple data items simultaneously. Further, the throughput of SIMD memory instructions must match SIMD compute throughput, or applications using SIMD execution will be bandwidth-bound: their performance will be limited by the bandwidth that a core can extract out of the memory system. In this chapter, we discuss SIMD memory instructions and their implementations for data laid out in both contiguous and non-contiguous patterns.

4.1 CONTIGUOUS PATTERNS

For computation with data parallelism, the most common memory access patterns are contiguous ones; e.g., in a loop with counter i that is incremented by one each iteration, we access all elements A[i], in order. Extending scalar load and store instructions to capture such patterns is relatively straightforward. SIMD load and store instructions simply read or write, respectively, a vector's worth of contiguous elements from a specified address—this is the same as scalar load or store instructions that can access a very large data item. To implement these instructions efficiently, we need only support wide reads and writes to the data cache. For accesses up to a certain size, this requires very little new hardware. Conventional data caches are organized in cache lines containing contiguous chunks of data to exploit spatial locality—whether an instruction accesses a small part of that line or a larger part of it makes little difference except for the number of wires we need between the cache and the core.

If the SIMD width is larger than a cache line, though, hardware will need to access at least two consecutive lines for a single SIMD memory instruction. To perform these accesses simultaneously, we would need significant changes to the cache controller and/or additional cache ports. An alternative is to spread the accesses across multiple cycles, which will reduce the throughput of SIMD memory instructions—e.g., if our SIMD width is twice the cache line size, and we access two lines across two cycles, our L1 bandwidth is half a vector per cycle. Yet another alternative is to increase the cache line size.

4.1.1 UNALIGNED ACCESSES

Even if our SIMD width is no larger than a cache line, one common scenario can trigger a need for a single SIMD read or write to access multiple cache lines: *unaligned accesses*. Unaligned accesses are not unique to SIMD loads and stores; any load or store of a chunk of data of size s is unaligned if the $\log_2 s$ least significant bits of the address are not zero. If the data item accessed straddles a cache line boundary, then hardware must access both cache lines to complete the operation—this is called a *cache line split*.

Example 4.1   Figure 4.1 shows three different unaligned memory accesses. Each chunk of memory is a full cache line, 64 bytes in size. The first example shows a four byte (i.e., scalar) unaligned access at address 0xbe, straddling a cache line boundary at 0xc0; this is a cache line split. The second example shows an unaligned SIMD access of 32 bytes that touches only one cache line, and the third example shows a 32B access that is a cache line split.

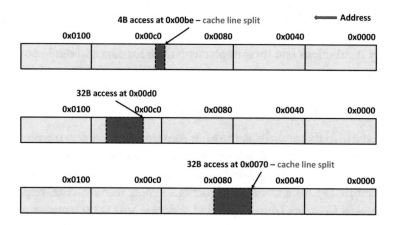

Figure 4.1: Example unaligned memory accesses.

Since cache line splits may occur on any architecture that allows unaligned memory accesses, regardless of cache line size, hardware must support them via one of the approaches described earlier, minus larger cache lines: simultaneous or serialized accesses to adjacent lines. The cost of adding support for simultaneous accesses to adjacent lines can be prohibitive, so architects often spread the accesses across multiple cycles. For reads, we can buffer the first line (or at least the desired portion) in a *split register*. Once we read the second line, we can concatenate the two lines and use a shifter to extract the desired unaligned data.

This approach means cache line splits execute at half the throughput of other cache accesses, and so may significantly degrade performance. However, for scalar data accesses, line splits are relatively rare for two reasons. First, most data is element-aligned; e.g., arrays of 32-bit integers are typically four byte aligned automatically by the compiler when it allocates memory. Second, even for unaligned data, scalar data items are much smaller than cache lines, so a random unaligned access is unlikely to straddle a cache line boundary. Consider an array of 32-bit integers on a system with 64 byte cache lines. If we iterate over contiguous elements in the array, every sixteenth access will be a cache line split; this will have relatively little performance impact.

For SIMD execution, we are much more likely to have unaligned accesses, and we "consume" cache lines much more rapidly. Even if a data array is element-aligned, we may want to access chunks that are not SIMD-aligned. For example, consider a computation where we add B[i+1] to A[i], for a set of contiguous values of i. If A[0] and B[0] both are SIMD-aligned, then our accesses to B will be unaligned. Also, in contrast to scalar execution, where a single access brings in a small portion of a cache line, SIMD accesses bring in a much larger part of a cache line, and possibly an entire cache line. Going back to our B[i+1] example, if our SIMD width is the same as a machine's cache line size, *every* access to B is a cache line split. Thus, to maximize the usage of the data cache's throughput, and also to enable vectorization on architectures without unaligned access support, a great deal of software optimization effort is spent on aligning data arrays [Bik et al., 2002, Eichenberger et al., 2004, Wu et al., 2005].

4.1.2 THROUGHPUT IMPLICATIONS

The high rate of data consumption of SIMD execution can affect performance even for aligned accesses. In particular, cache miss frequency sees a similar magnifying effect from SIMD as unaligned accesses. Consider a simple example of sweeping through an array of 32-bit integers on a system with 64 byte cache lines—we can fit 16 integers on a cache line. With scalar execution, due to spatial locality, we will miss in the cache only once every sixteen accesses. With a SIMD width of 512 bits, however, we will miss in cache *on every access*.

Latency from cache misses is typically not the biggest performance problem with SIMD execution, at least, not for contiguous access patterns. Prefetching may cover the latency of these cache misses—after all, we are considering contiguous accesses, which are likely to be captured by even a simple hardware prefetcher or by a compiler capable of inserting software prefetches.

Instead, bandwidth is the biggest performance challenge from SIMD memory accesses. SIMD execution increases bandwidth usage across all levels of cache, and all the way to main memory. With 100% SIMD efficiency, SIMD execution increases bandwidth usage by the vector length. Depending on the SIMD width and data type, that may be an order of magnitude increase. Lower SIMD efficiency does not necessarily mean lower bandwidth usage. Inefficiency from sources such as scalar instructions or "extra" SIMD arithmetic instructions may indeed reduce the bandwidth requirements of a SIMD computation. However, some vectorized computations include extra SIMD memory accesses, offsetting the bandwidth reduction from reduced SIMD efficiency.

One feature that can help mitigate bandwidth usage is streaming, or non-temporal, stores [Leverich et al., 2007]. Streaming stores are not specific to SIMD execution, but in at least some instruction sets they are linked. Streaming stores are store instructions that trigger no-write-allocate behavior in the caches. It may not be immediately obvious why this can save bandwidth. Let us consider a conventional store instruction that misses in the cache hierarchy. Hardware will read the cache line from memory, bring it into the cache hierarchy, and write the portion of the cache line corresponding to the store instruction. For a write-back cache, at some point, the line will be evicted and written back to DRAM. Thus, the store triggers both a read and write of the entire cache line. With a no-write-allocate policy, streaming stores can be performed directly in DRAM, and save the read of the line. This reduces bandwidth usage of writes by a factor of two, if we write the whole line, assuming the software does not access the cache line again before it would have been evicted from the cache.

If the SIMD width is smaller than a cache line, we may want to augment the hardware to capture spatial locality in streaming stores. Consider a loop where we write all elements in a cache line using streaming stores, but it takes us multiple streaming stores to do this because the SIMD width is smaller than a cache line. If we perform each of those streaming stores one at a time to DRAM, we may occupy the memory controller-to-DRAM channel *longer* than if we used conventional stores; thus, our effective bandwidth usage may be *higher*. However, we may be able to combine the streaming stores into a single full cache line write. Conventional write combining buffers already do this, so we may use the same technique for streaming stores. There are potential memory consistency model implications of this technique.

We have only discussed streaming stores, because while streaming loads also exist, their purpose is to avoid cache pollution (also a benefit of streaming stores). Reducing cache pollution can save bandwidth, but the effect is much less dramatic than the savings we see from eliminating memory reads with streaming stores.

Streaming stores perform writes all the way in main memory, and thus do not save bandwidth for accesses to data already in cache. This is a limitation; plenty of applications have working sets too big for the first level data cache, but small enough for other caches. We may thus extend streaming stores (and loads) to "stream" from a cache such as a second or third level cache, and capture the benefits of streaming accesses within the cache hierarchy [Park et al., 2013].

4.2   NON-CONTIGUOUS PATTERNS

While data parallel computations most commonly access memory in a contiguous pattern, plenty of computations access non-contiguous data items. Additionally, we must sometimes deal with unknown access patterns. There are algorithms where software cannot know at compile time whether data will be contiguous or not. In this situation, most programmers or compilers conservatively assume that data is non-contiguous and use non-contiguous SIMD memory operations—we discuss alternatives later.

Most non-contiguous memory accesses come from indirect array accesses or pointer dereferences, as illustrated in Figure 4.2. Array B contains offsets into array A, an array of integers; array ptr contains pointers into a set of integers. The computation performs indirect reads and writes to A, and indirect reads from ptr.

To vectorize the loop, we use *gather and scatter instructions*, sometimes known as indexed loads and stores, to perform these indirect accesses in SIMD fashion. These instructions access a set of memory locations specified by a base address and a set of indices. To use them, we first load a set of indices (e.g., elements of B or ptr) into a vector register, and then do a gather or scatter

```
// Scalar loop
for (i = 0; i < N; i++)
{
    A[B[i]] += *(ptr[i]);
}

// Vectorized loop with gather and scatter
for (i = 0; i < N; i += VLEN)
{
    vload v0, B[i]
    vload v1, ptr[i]
    vgather v2, A, v0
    vgather v3, 0, v1
    vadd v4, v2, v3
    vscatter A, v0, v4
}
```

Figure 4.2: Example loop resulting in non-contiguous memory accesses, and vectorized version with (simplified) gather and scatter instructions. ptr is an array of pointers.

by passing those indices and the base address of the array (e.g., A or 0 for pointer dereferences). The hardware scales each index by the size of the data type (not shown here for simplicity) and adds the result to the base address to produce a set of addresses. For a gather, it then reads that set of addresses and places the data into the output register. For a scatter, it takes in another input register holding a set of data elements, and writes those elements to the set of addresses.

Example 4.2   Figure 4.3 shows an example execution of the first gather instruction from our code example. The inputs are on the top of the figure, and the output on the bottom. The base address, the address of A, is 0x2000. The indices, from B, are in v0. Since A is an array of 32-bit integers, this gather must scale the indices by a factor of four before adding them to the base address, although this is not explicitly shown in the figure. This scale factor could be part of the opcode for the instruction, or could be another operand to the instruction (e.g., provided as an immediate value). The hardware reads the elements (A[20] at address 0x2050, A[0] at address 0x2000, ..., A[2] at address 0x2008) and places them into v2.

Example 4.3   Figure 4.4 shows an example execution of the scatter instruction from our code example. The base address and indices are the same as in our gather example, as we are scattering back to the same locations we just gathered from. The other input to the scatter is the set of values to write, in v4. The hardware writes the values from v4 to memory.

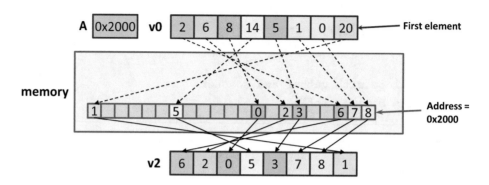

Figure 4.3: Example gather instruction. The inputs and output are on the top and bottom, respectively.

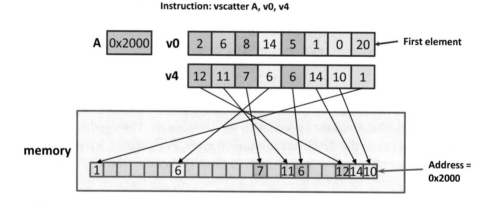

Figure 4.4: Example scatter instruction.

Gather and scatter instructions originated with processors such as the Cray-2 [Cray, 1989], Fujitsu's VP-100/200 [Miura and Uchida, 1983], and NEC's SX [Watanabe, 1987]. More recently, Intel has added gather and scatter to CPUs with short vector SIMD: AVX2 has gather instructions, Knights Corner has a limited form of gather and scatter instructions [Intel, 2014d], and AVX-512 has full gather and scatter instruction support [Intel, 2014c].

Non-contiguous SIMD memory operations are much more complex than contiguous ones, and so require much more hardware (and energy) and/or execute significantly slower (discussed in detail later). Thus, where possible, it is much more efficient for software to use contiguous memory operations.

To efficiently handle the case of an unknown access pattern, we can dynamically check for contiguous addresses. If done in software, the instruction overhead for this check is almost always going to be larger than the benefit. If done in hardware, the hardware needed is non-trivial, and this can add several cycles to the latency of a memory access. Some GPUs perform this check since the memory pipeline is already very deep [Rogers et al., 2013]. CPUs that rely on low latency access to a data cache are unlikely to be able to afford this hardware.

4.2.1   PROGRAMMING MODEL ISSUES

While we have described the basic functionality of gather and scatter instructions, there are a number of issues we need to address in a real system for software to be able to use them, and for architects to know the restrictions on implementation.

Completion Mask

Unlike load and store instructions, either scalar or SIMD, gather and scatter instructions have the ability to touch several pages. This has major implications on the definition and implementation of the instructions.

Let us first consider worst-case behavior for regular loads and stores. Unless a system supports scalar or SIMD data accesses larger than a page, a single scalar or SIMD load or store can touch at most two pages. To touch two pages, the access must be a *page split*, analogous to a cache line split, but straddling a page boundary. On a page fault or other exception triggered by accessing a page, processors supporting precise exceptions typically (1) roll back state such that it appears that the instruction *just before* the offending load or store was the last one executed, and (2) jump to the page fault handler so that when the handler completes, the offending load or store is re-executed. It is the fault handler's responsibility to remove the cause of the exception so that on re-execution the hardware will not trigger the same exception again. However, if the exception is triggered by a page split, we are not guaranteed that the offending load or store will succeed on the second attempt—the second page touched may also trigger an exception. Presumably, the fault handler will then clear the source of that exception, and the instruction will finally succeed on its third attempt. But what if the source of the first exception reappears? For example, on a system capable of holding only one page of physical memory, we may suffer a page fault for one of the two pages we need, swap in the page that triggered the fault, then suffer a page fault on the other page we need, etc. To guarantee that the application *eventually* makes forward progress, the operating system must guarantee that the hardware will eventually be able to access the two pages in a single execution of the instruction.

Now consider worst-case behavior for gathers and scatters. Gather and scatter instructions may access multiple elements, each of which may be a page split. If a gather or scatter instruction can touch N elements, it may touch $2N$ pages. Modern operating systems will generally not provide a guarantee on the number of pages an application can have in physical memory simultaneously beyond the two needed for page splits for regular loads and stores. Thus, we cannot guarantee forward progress of gather and scatter instructions under the same exception behavior we have for regular loads and stores. The fundamental issue is that we cannot guarantee forward progress of gathers and scatters if they happen in an all-or-nothing fashion—we must be able to make *partial progress* of gather and scatter instructions.

Making partial progress means allowing a gather or scatter instruction to modify architectural state even when it suffers an exception—hardware will read some subset of the elements and place them in the destination register, for gathers, or write a subset to memory, for scatters. On re-execution, however, how does the hardware know that a subset of elements has already been handled? Recall that this is a matter of making forward progress rather than preventing the hardware from repeating reads or writes; besides, repeating reads is presumably acceptable since it can happen on a cache line split, and repeating writes may be acceptable for some systems. Hardware must record in an architecturally visible place which subset of elements it has already handled, or at least, which elements it has left to handle. Recording this information as microarchitectural state (e.g., a register invisible to software) cannot provide any guarantees since that state may be lost on events like context switches. The most straightforward solution is to use an architectural register as a *completion mask* [Hughes et al., 2013].

A completion mask indicates which elements are yet to be read or written by a gather or scatter instruction. As with SIMD conditionals, for a completion mask, we can use either a SIMD register or a mask register. In this case, the mask is both an input and output. As an input, hardware examines the contents to determine which elements it should read or write. As an output, hardware clears the mask bits corresponding to elements it has successfully read or written. If a gather or scatter instruction is not interrupted, software can expect all elements to be read or written, and the mask to be completely cleared. Otherwise, software (including the same gather or scatter, on re-execution) can use the updated mask to see which elements still need to be read or written.

One side effect of updating the completion mask in this manner is that we cannot tell which elements were read or written by the instruction since hardware overwrites the input mask. A solution is to use separate input and output mask registers, but that presents several challenges. First, from a cost point of view, we need to encode another register in the instruction, which makes gather and scatter instructions larger. Second, this increases register pressure, which makes it more likely that we will need to spill mask registers. Finally, if we have separate input and output masks, we cannot simply re-execute a gather or scatter instruction after returning from a fault handler, since the input mask will not have changed. We would need to copy the output mask to the input, either in hardware (which completely ruins the purpose of separate input and output masks), or in software.

Finally, for gathers, since on a re-execution the hardware cannot know if it previously read any elements, the updates to the destination SIMD register must use blending masking. Otherwise, we may zero out elements that we have already read, and that the software wanted gathered.

Example 4.4   Figure 4.5 shows an example partial execution of the gather of $A[B[i]]$ from Figure 4.2. The top of the figure shows the inputs. The bottom of the figure shows the outputs, the mask register m0 and the destination register v2, both before and after the gather executes. The starting value of the mask indicates that all eight elements should be gathered. After reading five of the elements, the instruction is interrupted. The mask is updated to indicate the completed elements, and the values for those elements are written to the destination.

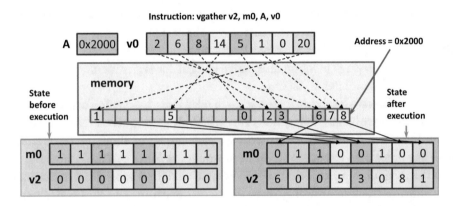

Figure 4.5: Example of a gather instruction that triggers a fault and only makes partial progress.

Ordering of Elements

Since a gather or scatter instruction is a read or write of multiple, potentially non-contiguous elements, architects must decide in what order the accesses may occur, so software developers can reason about the behavior. Relative ordering of loads and stores is governed by the memory consistency model. For certain models, we may not (visibly) reorder non-overlapping reads or writes within a single thread of execution. Other models are less restrictive and allow reordering of reads with respect to each other, writes with respect to each other, and/or reads and writes with respect to each other. For the most flexible models, gathers and scatters face no ordering constraints, but for others, we need to define the (apparent) inter-instruction and intra-instruction ordering of the reads and writes from gathers and scatters. This includes the ordering of overlapping and non-overlapping reads and writes in the inter- and intra-instruction cases. Two memory accesses overlap if they access one or more bytes in common.

We start with inter-instruction ordering, since this behavior is more intuitive. Here, the most obvious choice is to enforce the same ordering that is applied to other loads and stores.

Hardware cannot execute overlapping instructions out of program order to prevent violation of data dependences. For example, in a read-after-write situation where a scalar load comes after a scatter in program order, we would expect the load to return a value written by the scatter if any of the scattered elements overlap the load. For non-overlapping instructions, we can simply extend the memory model's restrictions to cover gathers and scatters.

We can extend the existing hardware for ordering to enforce the already-existing inter-instruction ordering restrictions on gathers and scatters. Most cores with out-of-order execution use load and store buffers to enforce both data dependences and any ordering restrictions on non-overlapping reads and writes. We can use the conventional load and store buffers for gathers and scatters as well—we need only allow a gather or scatter to allocate and fill multiple entries in these buffers rather than the usual one entry. The downside is that code with many gathers or scatters may use up the entries in these buffers very quickly, and that may stall the core. We may add more entries to compensate, but in conventional designs, the cost of these structures grows quickly with the number of entries, since they are typically content addressable memories. To make this more affordable, we may leverage work on alternative designs for larger memory ordering buffers that are much cheaper [Gandhi et al., 2005].

We could also save hardware by choosing to relax inter-instruction order for gathers or scatters. For example, if we believe that gathers are exclusively used to access read-only data structures, under an otherwise-strict consistency model we may choose to allow the reads from gathers to happen in any order with respect to stores in the program. Since this introduces significant complexity to the programming model, it should be motivated by a substantial performance boost.

For intra-instruction ordering, since the elements within a SIMD register are not inherently ordered, one could argue that any such restrictions are artificial. Further, performance way be improved if we allow hardware to read or write elements in no pre-determined order, as we will discuss later. However, in at least one case, the order that the hardware performs accesses is architecturally visible: scatters with overlapping elements. To help software to reason about SIMD memory behavior, e.g., to guarantee deterministic execution, we may want to place restrictions on at least this one case.

Example 4.5   Figure 4.6 shows a vector's worth of writes to A[B[i]] from our code example in Figure 4.2. One the far left, it shows the outcome of scalar execution. On the right side it shows the inputs to the scatter, and two possible outputs. On the left, it shows the outcome of SIMD execution where the scatter orders overlapping writes from first to last—this matches the scalar execution. On the right, it shows another ordering of writes that leads to a different result.

It can be quite useful to restrict overlapping writes from a scatter so that they appear to have happened from "first" to "last." This allows a vectorizing compiler to present the illusion of the original scalar execution order of loop iterations. Without this ordering guarantee, such a compiler must either guarantee no overlap among the writes in a scatter (e.g., via a dynamic check and/or regrouping of elements), or may have to avoid using scatter instructions. Happily, enforcing this ordering guarantee is not difficult, for at least some implementations.

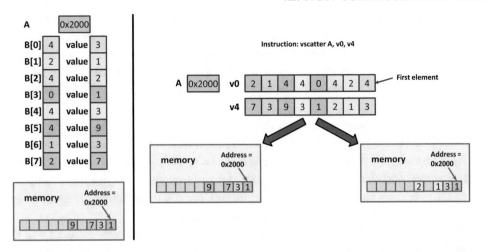

Figure 4.6: Scatter operation where we write eight values of A[B[i]]. The left portion shows scalar execution, where we individually write each value to memory. The right portion shows SIMD execution via a scatter instruction. For a scatter, we load B[i] into v0 and the values to scatter into v4. The output state, in memory, is shown for both sides, with two possible outputs shown for the scatter, depending on how overlapping elements are ordered.

If we extend memory ordering buffers to handle gathers and scatters by using one entry per read or write, intra-instruction ordering will inherit the inter-instruction restrictions. For example, if we cannot reorder loads in a certain memory model, then whatever order we place the reads for a single gather instruction into a load buffer is the order those reads will appear to happen. This mechanism will therefore provide the ordering guarantee we want for overlapping stores—we need only place the stores into the memory ordering buffer in the desired order. The downside is that in other cases, we may not *want* to enforce intra-instruction ordering. Thus, if this forces us to take on unwanted ordering restrictions, we way want to use an alternative ordering mechanism. If we are willing to live with whatever additional ordering restrictions are imposed by the memory ordering buffer, we may still not want those restrictions architecturally defined. That is, even if non-overlapping stores may be ordered in some way by the hardware, we may wish to hide that from software; this may provide microarchitects with at least *some* freedom in reordering execution.

Element Completion Guarantee

We've discussed gather and scatter ordering constraints and completion masks, and now address the interaction of the two. That is, when a gather or scatter is interrupted and we only make partial progress, what guarantees should we provide about which elements are completed? This element completion guarantee is *not* the same as the ordering of reads and writes. Instead, the completion guarantee describes what properties must hold regarding the state of the completion mask and destination register/memory when the instruction finishes (e.g., delivers an exception and jumps to the exception handler). The elements that *are* completed may be read or written in whatever order the hardware likes, within the element ordering restrictions. Poor choices about element completion may impact a forward progress guarantee or make gathers and scatters hard to use in practice.

A completion mask, or some equivalent, is necessary to guarantee forward progress of gather and scatter instructions, but it is not sufficient. If hardware has the freedom to complete elements in an inconsistent way, then we may still get stuck in an infinite loop. For example, suppose we execute a gather and this triggers an exception for the first element—we make no progress but jump to the exception handler. On re-execution, suppose the third element triggers an exception, and the hardware delivers that before completing any elements.

To avoid livelock in such a situation, we must guarantee that *some* element will make forward progress every time we re-execute a gather or scatter. The easiest way to do this is to pick some element, such as the "first" element (or closest to it with its mask enabled), and guarantee that either *it* will complete or the hardware will deliver an exception for *it*. That way, if the first element triggers an exception and the exception handler clears its cause, we will make forward progress on re-execution.

This does not prevent the hardware from making progress on other elements, unless this violates the element ordering restrictions. For example, assuming that we do not prevent reads from happening out of order, and the first element of a gather triggers a page fault, we may choose to complete the reads on all other elements before delivering the fault. If another element triggers an exception as well, we must suppress that exception, and make sure that we do not mark the element as complete when we finally deliver the exception on the first element.

Traps and interrupts create a problem with this approach. Like exceptions, they are detected during execution, but unlike exceptions, they do not prevent completion of an instruction. Thus, they are typically delivered *after* an instruction commits.

If a gather or scatter triggers a trap, or an interrupt occurs during its execution, but no exceptions are triggered, then we can complete the instruction and deliver the trap(s) and/or interrupt(s). However, we face a conundrum if one element triggers a trap, such as a data breakpoint, and another triggers an exception. For normal loads and stores that trigger both a breakpoint and exception, we deliver the exception, discard the breakpoint trap, and on re-execution, we will (presumably) re-trigger the breakpoint. However, for gathers and scatters, we make partial progress. If we allow the element that triggered the breakpoint to complete, and deliver the exception, then the information that we hit the breakpoint may be lost forever. That is unacceptable since it would render traps and interrupts completely unreliable in applications using gathers and scatters.

One solution to this is when we only make partial progress, check for traps or interrupts triggered by completed elements, and if there are any, deliver those instead of the exception [Hughes et al., 2013]. This way we do not lose the trap or interrupt information. Further, there is no harm in dropping the information that we hit an exception because the element that triggered it is not marked complete; when we re-execute the instruction, that exception will most likely be re-triggered. However, this may change the model for traps and interrupts—the handlers for these must return to the instruction that triggered them, if that instruction is not yet complete.

4.2.2   IMPLEMENTING GATHER AND SCATTER INSTRUCTIONS

Before we discuss how to implement gather and scatter instructions, let us first think about what each of these complex instructions really does:

- splits a vector of indices and mask bits into individual elements;

- for a scatter, also splits a vector of data items into individual elements;

- for each element, scales the index, adds it to the base address, and uses the mask bit to conditionally fire a read or write;

- for a gather, waits for each element to come back from memory, and merges it into the appropriate place in the destination register; and

- zeros out the completion mask

This is far more work than that done by a SIMD load or store instruction, and more even than performing a scalar load or store for each data element. Further, this list does not include the work that must be done to ensure proper ordering of elements, as discussed in Section 4.2.1. Therefore, building a high performance gather and scatter implementation is extremely challenging.

We now describe considerations for hardware support in the core and memory system for gathers and scatters.

Hardware in the Core

Within the core, we have a spectrum of hardware support to choose from when implementing gather and scatter instructions. On one end, we may use existing scalar and SIMD operations, and add little hardware to a core. On the other end, we may add dedicated hardware to generate addresses in SIMD fashion, and add a state machine to create, track, and merge the results of (for gathers) the independent memory requests.

Ignoring the zeroing of the mask and memory disambiguation (discussed earlier), we have three stages of execution for a gather or scatter. For gathers, we must (1) generate the addresses, (2) perform the reads, and (3) merge the results. For scatters, we must (1) generate the addresses, (2) extract the data items to be written, and (3) perform the writes. For each of these stages, we may use software, or we may provide dedicated hardware.

First, let us consider a pure software sequence to perform a gather operation. Our sequence uses scalar conditional loads, and so address generation (index extraction) for each element takes one instruction, and performing the read takes two more, one to extract the mask bit and one for the load. Finally, merging each result takes another instruction (excepting the very first element). In total, including zeroing the mask, we need four times the vector length number of instructions to emulate a gather instruction. Roughly three quarters of the instructions are for unpacking and repacking information from/into SIMD registers, and only a quarter are the loads. The emulation sequence for a scatter instruction is similar: we can skip the merging step, but we need to additionally extract the data items to be written to memory.

The performance problem with a software sequence for a gather or scatter is not just the latency or throughput of the sequence, but the impact to other instructions in the application—the emulation sequence will compete with computation instructions for pipeline slots and functional units. Figure 4.7 shows the SIMD efficiency for a computation comprising a set of emulated gathers and SIMD arithmetic instructions. The efficiency calculation here assumes that the corresponding scalar code uses a single load instruction per gathered element. While the arithmetic instructions have 100% SIMD efficiency, the gathers have SIMD efficiency of $\frac{1}{4 \times VLEN}$. Thus, efficiency of the overall SIMD execution drops as SIMD width increases. Even with a vector length of only two, we need 62 arithmetic instructions per gather to hit 90% SIMD efficiency. Without hardware support, gathers and scatters can drown out performance benefits from SIMD execution.

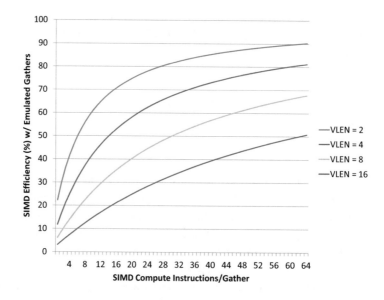

Figure 4.7: SIMD efficiency of a computation with a set of (emulated) gathers and SIMD compute instructions.

Now let us consider hardware to accelerate execution and/or offload the work from the main core pipeline to make room for other instructions.

We may generate all of the addresses in SIMD fashion, as shown in Figure 4.8. This hardware shifts each index by the scale amount, and then adds the base address to the shifter outputs to produce the full set of addresses. If the data cache, or whatever structure we use for gathers and scatters, supports enough independent accesses, we can then issue all of the reads or writes simultaneously. This requires additional hardware to extract all of the mask bits in parallel and pass them to the memory system with the addresses.

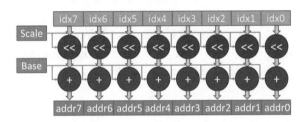

Figure 4.8: SIMD address generation for a gather or scatter.

Most data caches (and TLBs) cannot support so many simultaneous independent accesses. In fact, the number of memory accesses we can make at once often limits the throughput of gathers and scatters. If we cannot send all requests to memory at once, we need to buffer the addresses, or instead use hardware that only generates as many addresses per cycle as we have data cache ports.

Example 4.6   We can complete a gather or scatter once every $\frac{VLEN}{\#\ accesses/cycle}$ cycles. Suppose our vector length is 16, and our data cache has two ports. Unless we somehow combine accesses, at best we can complete a gather or scatter every eight cycles. Thus, if $\frac{SIMD\ arithmetic\ instructions\ per\ gather\ or\ scatter}{SIMD\ arithmetic\ execution\ rate}$ is less than eight, then performance will be limited by gathers and scatters, regardless of the hardware support for gathers and scatters. Further, most computations that use gathers and scatters also include other memory accesses. Those will compete for the ports, and reduce throughput further.

Some cores have dedicated address generation hardware tightly integrated with the data cache ports that is hard to bypass. In such cores, we cannot easily use SIMD address generation hardware. However, we can still provide hardware to extract and buffer individual mask bits and indices, and a state machine to pass them, along with the base address and scale, to the memory system for the reads or writes.

The remaining stages we're considering for hardware acceleration are result merging, for gathers, and data extraction, for scatters. To merge results into the destination as they return from memory, we need a buffer to hold the partial results, and a multiplexer to place each result into the proper position. To control the multiplexer, we either need each data element to carry its position in the output, or we need to retrieve that information from a structure holding a mapping of address to position in the output. Extracting data elements for scatters is the same as extracting indices and mask bits. If we use SIMD address generation and perform all writes simultaneously, then we need only split the data items and send each with its address to the memory system. Otherwise, we may need to buffer the individual data items.

Combining Elements

A gather or scatter instruction may exhibit spatial locality that hardware can leverage for improved performance. More specifically, multiple elements from a single gather or scatter may fall on the same cache line. If we access entire cache lines (or large pieces of them) from memory instead of individual elements, we may capture all elements with fewer accesses. Some architects refer to this as memory coalescing [Rogers et al., 2013]. Since throughput of gathers and scatters can be proportional to the number of elements we can read/write from/to memory each cycle, this may improve performance significantly.

Example 4.7 Figure 4.9 shows example indices for a gather or scatter instruction that is executed with cache line accesses. For this system, four elements fit on a cache line, and we can perform only one access per cycle. The hardware handles elements from right to left. In the first cycle, we access the first element, with index 73; the instruction has a duplicate index later in the vector which is also covered by this access. In the second cycle, we access index 13, which will touch a cache line with later index 12 as well. In the same manner, we access indices 7 and 5 in the third cycle. Finally, we access indices 51 and 24 by themselves in the fourth and fifth cycles, respectively. Thus, we complete our accesses in five cycles, when it would have taken eight otherwise.

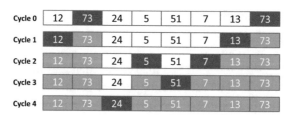

Figure 4.9: Indices for a gather or scatter, colored to indicate which have been accessed. Dark green indices are on the cache line accessed this cycle. Light green indices have been previously accessed.

A somewhat extreme example of spatial locality within a gather or scatter is one with a set of contiguous indices. This can result from software not being able to guarantee that the indices would be contiguous. We may provide hardware to automatically detect this case and convert it to a conventional SIMD load or store. However, such hardware is not cheap—we would need to detect that each index was the previous index plus one. If we have support to combine accesses on the same cache line for gathers and scatters, that may provide enough performance for this case.

Detecting and taking advantage of spatial locality within a gather or scatter requires significant hardware. The most straightforward technique to detect spatial locality, after computing the addresses of all elements, is to pick an element to access and compare its cache line address (i.e., the address of the beginning of the cache line on which it sits) to all other elements'. For gather/scatter implementations which do not generate all addresses up front, we can use partial addresses; e.g., the elements share the base address, so we need only use the lower-order bits of that.

If that comparison logic is too expensive, we may reduce its cost by imposing restrictions on which elements we may combine into a single memory access. For example, we may only compare the partial addresses of nearby or even adjacent elements in the vector.

Another cost is in extracting the desired elements out of each cache line (for gathers), or inserting them into the block written to memory (for scatters). Let us consider gathers. We want to read a whole cache line from memory, grab an element we want, and place it into the right position in the output vector. Further, we want to be able to do this for multiple, potentially all, output elements simultaneously. This requires a crossbar to allow any part of the input cache line to be routed to any/all of the output elements. Scatters require a similar network, but data flows from the SIMD data register to a buffer holding the elements on the same cache line to be written to memory. Scatter hardware must additionally be able to handle overlapping elements, especially with regard to any ordering constraints they may have—e.g., "later" elements must appear to overwrite "earlier" overlapping elements.

Like detection of spatial locality, the hardware required for full flexibility in extracting/inserting elements on a cache line is potentially expensive, but can be made cheaper with some restrictions. In particular, we may restrict the number of elements that can be combined into a single memory access, where those elements are in the vector relative to one another, and/or the number of requests of the *same* element we can combine into one request.

Another cost of taking advantage of spatial locality in a gather or scatter is unpredictability. Software and hardware both cannot know until execution time how many memory accesses a given gather or scatter instruction requires. Thus, instruction schedulers either need to predict this and be prepared to handle mispredictions; assume worst-case behavior, limiting the benefit of the optimization; or allow for some delay between detection of impending completion and scheduling a dependent instruction.

Memory System Support

For the memory system, just as in the core, we may choose to add or modify little for gathers and scatters, or we may add significant dedicated hardware to maximize performance of gathers and scatters. We focus on the most critical level of the memory hierarchy: the first level.

In most designs, the most dramatic change that can be made is to increase bandwidth for non-contiguous accesses. As discussed in Section 2.4, increasing bandwidth for contiguous accesses is relatively easy, to a point, since we just need to widen the data path from the cache to the core. In contrast, increasing the number of simultaneous independent accesses we can make to a cache is much more challenging. However, this can have a dramatic effect on performance of gathers and scatters.

Example 4.8   Assume our core can execute one SIMD arithmetic instruction per cycle, and begin up to one gather or scatter per cycle. Also assume that our computation has enough independent work that we do not stall on a data dependence, and that scalar computation is not on the critical path. Let a be the number of arithmetic instructions per gather or scatter. The steady state performance, in terms of SIMD arithmetic instructions executed per cycle, is $\frac{a}{\max(a, \lceil \frac{vector\ length}{elements\ per\ cycle\ from\ memory} \rceil)}$. If we have one gather or scatter per SIMD arithmetic instruction, performance is proportional to the number of elements we can read or write per cycle, up to the vector length.

To increase the number of elements we read or write from the first-level cache per cycle, the simplest way, conceptually, is to increase the number of read/write ports. However, this is very expensive, so we look for cheaper alternatives. First, though, we consider address translation.

We must translate the virtual addresses of all elements, and this translation rate must match the data access rate or it will be the bottleneck; thus, we face the same issue for the TLB: either increase the number of ports, or find a cheaper alternative. Since most designs do not worry about coherence in TLBs, replicating the TLB is one choice, and is much simpler than replicating the data cache [Espasa et al., 2002]. We may leverage a similar spatial locality trick for address translation that we described earlier for combining accesses to data elements on the same cache line. Since pages are typically much bigger than cache lines, we would expect most gather or scatter instructions to access fewer pages than cache lines. Another option is to have gather or scatter instructions skip address translation entirely. In most systems, this would require that the addresses used are in a different address space than regular loads and stores, such as those for a dedicated software-managed buffer. While CPUs have largely resisted such programming model changes, GPUs have embraced them, e.g., shared memory in CUDA [Nvidia, 2014] and local memory in OpenCL [Khronos, 2013].

To increase the (non-contiguous) access rate of the cache, in terms of elements per cycle, we have two separate problems to solve: (1) increasing the number of data elements we can pass back and forth between the core and cache, and (2) increasing the number of data elements we can read or write from the data array itself. For both, we can take advantage of elements being small. For (1), the data path between the core and cache is presumably large enough to support a full SIMD register; otherwise, we cannot support even a single SIMD load or store per cycle. Thus, even if elements are not on the same cache line, they may be packed together and shipped to or from the cache in (potentially) a single access. The address path, on the other hand, needs to be widened to support multiple accesses per cycle. For (2), if the data array is organized into multiple, independently controlled banks, then we may simultaneously access different cache lines on different banks.

Figure 4.10 shows an example data array from a cache that supports up to four element accesses per cycle. Each cache line is striped across the banks. That is, each line is broken into four chunks that can be independently accessed. The tag array supplies a row index for each bank, which will be the same when performing a conventional SIMD load or store. For gathers and scatters, we enhance the hardware to allow multiple tag lookups each cycle, one per bank.

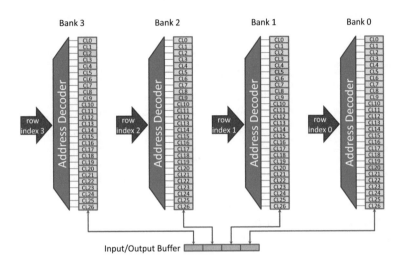

Figure 4.10: Data array for a banked cache, where each cache line (e.g., CL0) is split across the four banks. To access a single line, all row indices are the same. For gathers and scatters, we allow the row indices to be different. The orange entries indicate the accesses for an example gather or scatter.

For gathers and scatters, we may want to access chunks from different cache lines. Since the banks are independently controlled, we may access chunks on up to four different cache lines, as long as the chunks are in different banks. If two desired chunks are in the same bank, this is known as a *bank conflict*. When we have a bank conflict, we must spread the conflicting accesses across multiple cycles, unless a bank supports multiple accesses per cycle (i.e., has multiple ports).

Whether two accesses result in a bank conflict depends on their offsets into their corresponding cache lines. These offsets are typically the least significant bits of the addresses. The core knows these offsets before it sends the requests to the cache, and it can (and must) detect bank conflicts and send the necessary number of requests to the cache. In particular, at each cycle, it must pass a tag for each bank it wants to access, and it can only send one tag per port per cycle.

The number of cycles a gather or scatter occupies the cache is thus the maximum number of accesses the gather or scatter makes to a single bank, which is at least $\lceil \frac{vector\ length}{\#\ banks} \rceil$ and at most the vector length. Figure 4.11 shows the expected occupancy for a single gather or scatter instruction

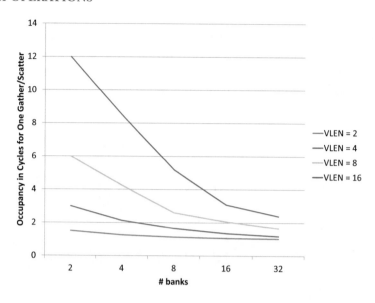

Figure 4.11: Occupancy of a banked cache for a single gather or scatter instruction as we vary the number of banks and the vector length.

as we vary the vector length and number of banks. This assumes a single cache port and that all banks are accessed with uniform probability. This also assumes that if the vector length is greater than the number of banks, each vector is split into chunks with as many elements as there are banks, and that we must complete all accesses for one chunk before moving onto the next one. Bank conflicts make it very difficult to achieve a throughput of one full vector per cycle, especially as vector length increases [Gou and Gaydadjiev, 2011]. With longer vectors, the odds of at least one bank conflict are quite high; e.g., for a vector length of 16 and 32 banks, we have a 99% chance of having at least one bank conflict.

In practice, some applications will access data in a strided manner, which can increase bank conflicts beyond what we'd expect from a uniformly random distribution of addresses. An application that is aware of this may use padding to avoid it, or hardware may shuffle the data from each cache line to avoid this pathological case.

As mentioned earlier, sending independent requests to the banks this way requires multiple tag lookups per cycle. This may be done either by adding ports to the tag structure, or by replicating the tags per bank.

An additional drawback of this approach is that the extra hardware in the cache, plus the bank conflict detection hardware, may increase the latency of memory accesses, and thus degrade performance. For bank conflict detection and minimization, we can control this somewhat via the level of sophistication of the hardware. In particular, we may choose to use simple access scheduling, as described above. An alternative is to reduce conflicts by reordering elements, assuming that is allowed by the programming model; the cost is increased hardware complexity, and possibly increased latency.

An alternative banked cache design interleaves entire cache lines between banks, rather than striping each line across the banks. For example, cache line 0 is in bank 0, cache line 1 is in bank 1, etc. This can be more efficient than the striping approach, since we need only active a single bank for many cache accesses. However, for such a design to allow high throughput for contiguous accesses, each bank must be capable of reading/writing the entire input/output buffer. Thus, for non-contiguous accesses, we either can activate only one bank per cycle, or we need (potentially complex) arbitration logic to decide which bank will read/write each data element. Further, bank conflicts are still an issue in such a design, since we may find ourselves wanting to access multiple cache lines in the same bank. Tarantula used such a cache design for its L2 cache, and included access scheduling logic to avoid bank conflicts [Espasa et al., 2002].

If modifying the data cache to support high throughput gathers and scatters is not an option, we may consider adding a dedicated data buffer for these instructions, i.e., a separate cache or even a software-managed buffer. For example, a software-managed buffer can eliminate several costs such as tag arrays and TLB ports. However, a new structure to improve gather or scatter throughput comes at considerable cost. The structure itself will consume significant area and power. It will also complicate software—software may have to explicitly indicate which memory structure to use, or at least be optimized to make best use of the new structure. Finally, depending on the programming model, we may need to expose the structure to the coherence mechanism, to ensure we maintain coherence between the different first-level structures.

4.2.3   LOCALITY IN GATHERS AND SCATTERS

One might assume that gathers and scatters in real applications exhibit very little spatial or temporal locality; after all, they can be used to access seemingly unrelated addresses from arbitrary data structures. However, just because the operations *enable* software to access large sets of pseudo-random locations, doesn't mean all software wants or needs to do this.

Computations With No Locality
Certainly, there are important computations that rely on gathers or scatters with no locality. For example, if we use SIMD to perform lookups in a large hash table or traversals through a large binary search tree, we will see little spatial or temporal locality unless the inputs are biased. Such access patterns place huge stress on the memory system. With neither spatial nor temporal locality, most data element accesses will go to main memory and will bring back an entire cache line; this creates a tremendous bandwidth requirement if gathers and scatters are common. It also causes massive cache pollution, since we'll touch so many distinct cache lines only once.

We may mitigate both of these problems. One option is sector caches [Liptay, 1968]. These break each cache line into chunks, and keep separate state information (e.g., valid bits) for each chunk. This allows us to transfer data in sub-cache-line sized pieces, reducing waste when we have little spatial locality. It is helpful to have a predictor to determine how much of each line to retrieve [Rhu et al., 2013]. Another option that can help if we have no spatial or temporal locality is to bypass the cache hierarchy entirely. Here, we may request as small a chunk as possible from main memory, and send it directly to/from the core. This is analogous to streaming loads and stores, discussed in Section 4.1.2.

Computations With Locality

Many computations needing gather or scatter *do* exhibit significant locality.

To exhibit temporal locality and still require gathers and scatters, an access pattern need only be *likely* to re-use some data within a short time of first touching it. One application behavior that leads to this is indirect accesses to small data structures, such as a small lookup table of which we access input-dependent entries, e.g., for function approximation [Wawrzynek et al., 1996]. Another extremely common pattern is where, in an inner loop, we iterate over a set of indirectly accessed objects, and that set changes slowly over outer loop iterations. This can provide significant temporal locality, even if the total data structure is very large. Three concrete examples of this are as follows.

First, most sparse matrices have structure to their matrix because they are based on a physical system. The result is that adjacent rows in the matrix have non-zeros in many of the same columns. For key sparse linear algebra operations like sparse matrix-dense vector multiplication, we typically use SIMD to process multiple elements in a row of the matrix at a time. Further, the elements in the vector that we need are non-contiguous, since they correspond to the columns of the non-zeros in the row; thus, we use gathers and scatters to access them. As we iterate over the rows in the matrix, we re-use many of the same elements from the vector.

Second, in the molecular dynamics example from Section 1.2, we use SIMD to operate on multiple neighbors of each particle. The neighbors change over time, so it is hard to keep them contiguous; thus, we use gathers to read information about the neighbors, such as their positions in space. Nearby particles in space will have many common neighbors; thus, as we iterate over the particles, we re-use much of the information from previous particles' neighbors.

Finally, in the speech recognition example from Section 1.2, we process the active states, or their neighbors (in some approaches), in SIMD fashion. The set of states we access is non-contiguous, and therefore we need gathers and scatters to do so. As we process speech fragments, the set of active states changes, but slowly; thus, across speech fragments, we re-use much of the state information.

There are a few ways in which a gather or scatter can exhibit spatial locality. First, it may be used to access a small data set, where the chance of touching multiple elements on a cache line is inherently higher. Second, it may be used to access a large fraction of elements in a data set, and the indices may be ordered. For instance, we may iterate over all or most of the elements in an array, but we may do so in a pseudo-random order. In cases like this, software may even be able to reorder accesses to increase spatial locality; for example, it may be able to sort or partially sort the index array. Finally, it may be used to access data that naturally has some spatial locality. For instance, in our sparse matrix example above, not only do adjacent rows often exhibit similar indices, but the indices themselves are often clustered, e.g., around the diagonal of the matrix.

Another extremely common way gathers and scatters can exhibit spatial locality is *across* multiple gathers and scatters. Figure 4.12 shows a canonical example of this. The input data is kept as an array of structures (AoS), with four fields in each structure. This is common, for example, when we represent an array of objects with multiple properties such as coordinates in space. The output data is kept as a structure of arrays (SoA), with four separate arrays that each hold all of the elements from one of the fields of the AoS.

```
// Scalar loop
for (i = 0; i < N; i++)
{
    X[i] = A[B[i]].x;
    Y[i] = A[B[i]].y;
    Z[i] = A[B[i]].z;
    W[i] = A[B[i]].w;
}


// Vectorized loop with gathers
for (i = 0; i < N; i += VLEN)
{
    vload v0, B[i]
    vmul v0, v0, 4
    vgather v1, A, v0
    vgather v2, A+4, v0
    vgather v3, A+8, v0
    vgather v4, A+12, v0
    vstore X, v1
    vstore Y, v2
    vstore Z, v3
    vstore W, v4
}
```

Figure 4.12: Example of AoS-to-SoA conversion.

The vectorized loop to convert the data from AoS to SoA format uses one gather instruction for each field of the structures. It multiplies the indices (into the AoS) by the number of data elements in each structure (four), to gather a set of like elements. That is, it gathers a set of x, then a set of y, etc. After that, it uses SIMD stores to perform contiguous writes to the SoA output.

If we look more closely at the gathers, we see that they all use the same indices, but their base addresses are offset by the size of one data element (four bytes, in this case). The result is that the second gather touches elements adjacent to those touched by the first gather, and so on, to the end of the loop iteration.

Example 4.9   Figure 4.13 shows the elements touched by each gather in our AoS-to-SoA conversion. This example assumes a vector length of two, and that the elements of B for this SIMD iteration are zero and seven. The first gather reads x0 and x7, the second y0 and y7, etc. If each structure takes up a whole cache line, then this sequence only touches touches two cache lines, but repeatedly.

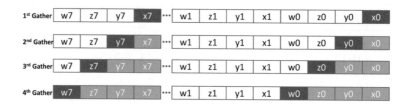

Figure 4.13: Gather behavior for AoS-to-SoA conversion. Dark green elements are accessed by this gather. Light green elements have been previously accessed.

Gather and scatter hardware can take advantage of this cross-instruction locality [Batten et al., 2004, Forsyth et al., 2014]. One option is to define gather and scatter instructions that explicitly perform accesses to data in AoS format. The underlying hardware can read or write chunks of data containing multiple elements. Another option is for hardware to predict that data might be in AoS format and buffer it to reduce the number of accesses to the memory system. For example, we might buffer all of the lines accessed by the first gather in our example, and the remaining gathers may read from that buffer at higher performance and lower power than if they had to read from the first-level cache.

CHAPTER 5

Horizontal Operations

Thus far, we have discussed applying SIMD execution to completely data parallel computations. This allows hardware to treat vector elements as completely independent of each other. However, many computations with data parallelism also contain some dependences between computations on different data items (i.e., elements in a vector). This requires a mechanism to pass information from one element in a vector to another, or to combine (i.e., reduce) elements.

A naive way to do this is through memory—with unaligned loads and stores, and gathers and scatters, we can write a vector to memory and then read it back into a SIMD register, shifted or scrambled. This can be expensive in terms of performance, since we need to either wait for the write to finish before reading, or we need to forward the data from the store instruction(s) to the load instruction(s), which is often a slow path. Thus, a mechanism to perform inter-element SIMD operations directly on one or more SIMD registers can provide a performance boost.

SIMD operations that move or combine data between elements in the same vector are known as *horizontal operations*. In this chapter, we discuss horizontal data movement instructions, reduction instructions, using horizontal instructions to help with control divergence, and instructions and techniques to handle dependences between elements in a vector that can only be detected dynamically. First, though, we look at how horizontal operations scale with SIMD width.

5.1 LIMITS TO HORIZONTAL OPERATIONS

Horizontal operations are special because of their ability to move data from one position in a vector to another. Some perform arithmetic on the resulting rearranged data, and some simply pass the rearranged data onto the output. Horizontal operations that perform arithmetic are essentially the same as other SIMD arithmetic instructions, except for the data rearrangement. The costs and tradeoffs for horizontal operations are dominated by data rearrangement.

During the rearrangement phase of execution, hardware selects for each output element e_i, an input value v_j from one of its input vectors, where i and j do not have to be equal. A fully general operation allows software to choose any j for any i, and would further support an arbitrary number of input vectors. In practice, the more flexible the operation, the more expensive and/or slower it is.

In particular, brute force hardware to rearrange elements in an arbitrary way within a vector has *quadratic cost* with the length of the vector. As with arithmetic hardware, for vector length N, we need N pieces of logic (e.g., a multiplexer) to produce one output value each. However, unlike conventional SIMD arithmetic operations, each of those pieces of logic has a number of inputs proportional to N—all v_j must be inputs for all output elements, or we cannot produce arbitrary data rearrangements. With k input vectors, we effectively have an input vector length of kN for each output element, or a total of $(kN)^2$ inputs to produce N outputs.

This unscalable cost means hardware to rearrange data arbitrarily is expensive for even modest vector lengths. We may use a multi-stage switching "network" to reduce the logic complexity and number of wires, but this comes at higher latency.

Architects therefore sometimes provide *restricted* flavors of data rearrangement, even if they also provide arbitrary (and slow) rearrangement. For example, instruction sets may include operations that only support moving each element to a neighboring spot in the vector (e.g., Krashinsky et al. [2004]), or may split a vector into groups of contiguous elements and only support movement within a group (e.g., Intel AVX).

5.2   DATA MOVEMENT

Data movement operations perform no arithmetic on their inputs, but rather select an input element for each output element. Basic data movement operations may have as input a single value (a *broadcast*), a single vector (a *permute*), or multiple vectors (a *shuffle*). Broadcast operations replicate a single input value across all output elements. Permute operations allow software to choose any of the input values in a single vector for each of the output values. Shuffle operations are similar, but take two (or more) input vectors.

Example 5.1   Figure 5.1 shows two hypothetical permute operations, where the lines represent the choices for each output element. The top one is unrestricted, in that software may select any input value for any output value. The bottom one is restricted; it divides the vectors into two groups of four elements, and software may only select an input element from the same group as the output element.

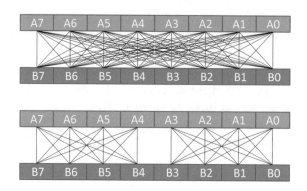

Figure 5.1: Unrestricted and restricted permute operations on A.

Example 5.2   Figure 5.2 shows two example shuffle operations. The top one conceptually concatenates two vectors and then chooses a contiguous set of elements to output. This sort of operation is very useful, e.g., for sliding window operations like filters. The bottom example in 5.2 shows a shuffle that conceptually concatenates two vectors and then outputs the "even" elements. This operation, also known as a strided load if done from memory, can be used, e.g., to extract one field out of an AoS. Figures 5.3 and 5.4 show example loops for both flavors of shuffle. The `vwindow` instruction performs the first kind of shuffle, and the `veven` instruction performs the second.

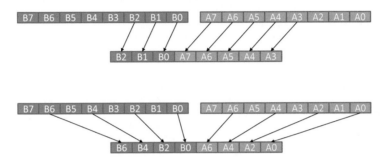

Figure 5.2: Example shuffle operations on A and B.

We have several choices for how to give software control over a data movement operation.

- Built-in: The instruction definition includes exactly how data will be moved. The `veven` instruction above is an example of this.

- Scalar value: The instruction takes a scalar value as a parameter. In some cases, a single value is enough to describe the operation of the instruction. The `vwindow` instruction above is an example of this. In other cases, the operation could use independent control values for each element, but we only allow software to provide one value, either to make the implementation of the operation simpler, or to force the control to be compact. The instruction may take this parameter as an immediate or in a scalar register.

- Vector of values: The instruction takes a vector of values as a parameter. With this option, software can specify a control value for each element, and thus has a lot of flexibility. The instruction may take this parameter as an immediate, in a scalar register, or in a SIMD register. For the first two options, we may have too few bits to specify all possible choices for each control value. To pack the control information into the limited space, we can either limit the set of choices for each control value, or we can force each value to control multiple elements.

Data movement operations are most often used to reorganize data structures or to reorder elements in a vector to satisfy data dependences. Another use is to implement very efficient lookup

```
// Scalar loop
for (i = 0; i < N; i++)
{
    for (j = 0; j < W; j++)
    {
        A[i] *= B[i + j];
    }
}


// Vectorized loop
for (i = 0; i < N; i+=VLEN)
{
    vload v0, A[i]
    vload v1, B[i]
    vload v2, B[i+VLEN]

    for (j = 0; j < W; j++) // W <= VLEN
    {
        vwindow v3, v1, v2, j
        vmul v0, v0, v3
    }

    vstore A[i], v0
}
```

Figure 5.3: Example sliding window computation.

tables. For example, if we can fit an entire lookup table into a SIMD register, and place the indices of the desired elements into another, then a permute operation will perform a SIMD table lookup [Diefendorff et al., 2000]. This is analogous to performing a gather, but out of a SIMD register instead of memory, and is therefore significantly more efficient.

Two final data movement operations, *compress* and *expand*, are fundamentally different from the operations described above. These operations are used to pack or unpack a selected set of data items from/to a vector. Unlike other horizontal data movement operations, where output elements are independent, compress and expand operations require implicit communication between different vector elements, to determine the right placement of the output values into the output vector.

Figure 5.5 shows an example compress operation, and its opposite, an *expand* operation. The register v0 contains a set of values, and m0 indicates which of those values we would like to

```
// Scalar loop
for (i = 0; i < N; i++)
{
    C[i] = B[2 * i];
}

// Vectorized loop
for (i = 0; i < N; i+=VLEN)
{
    vload v1, B[2 * i]
    vload v2, B[2 * i + VLEN]
    veven v3, v1, v2
    vstore C[i], v3
}
```

Figure 5.4: Example strided load computation.

keep. The compress operation packs the four desired values into the low part of the output register v1. The expand operation "unpacks" the compressed values from v1 into the elements of v2 whose mask bits (in m1) are set. Compress and expand from/to memory was supported by the Fujitsu VP-100/200 [Miura and Uchida, 1983], and the NEC SX vector processors extended this to include compress and expand directly in vector registers [Watanabe, 1987]. Intel's Knights Corner coprocessor brought memory compress and expand to short vector instruction sets [Rahman, 2012], and Intel's AVX-512 extends that to register compress and expand [Intel, 2014c].

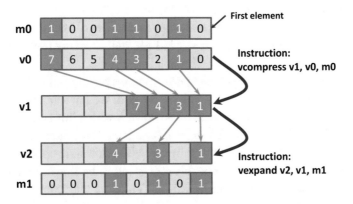

Figure 5.5: Example compress and expand instructions.

5.3 REDUCTIONS

Reduction operations take a set of values and combine some or all of them to produce a smaller set of values. For single-threaded code, software most often uses reductions to combine a full set of elements, e.g., to compute a sum of an array of values.

Let's consider how we would use SIMD execution to compute a scalar sum over an array. (Other common reductions, such as maximum, minimum, and product, use the same procedure.) The most straightforward way is to split the array into vectors, create a SIMD accumulator initialized to all zeros, and iteratively add each vector to the accumulator. At the end of this process, we are left with a SIMD register whose contents we want to add together.

One option for this final step is to use scalar execution. For example, we could use scalar code to read one value at a time from the SIMD accumulator and add it to a scalar accumulator.

Another option is to use SIMD execution, albeit somewhat inefficiently. The most efficient way to perform this reduction with SIMD execution is in binary tree-fashion. For instance, we first combine adjacent elements in the vector, then partial sums two elements apart, then four elements apart, etc. until we have a single element with the final result. Figure 5.6 shows an example of this technique, a *SIMD tree reduction*.

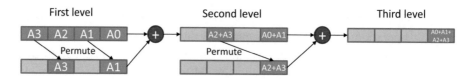

Figure 5.6: Example SIMD tree reduction. The light elements in each SIMD register hold "do not care" values.

This method computes the answer in $\log_2 vector\ length$ steps, rather than the $vector\ length - 1$ steps needed for the scalar approach. The SIMD tree reduction is not 100% efficient since as we proceed down the tree, we have half as many useful SIMD values as the previous level of the tree.

For each level of the tree, we actually perform two operations. First, we have a data movement operation to align two "nodes" in the tree. Second, we have an arithmetic operation—an addition, in our example. These operations are serialized, which limits what we as architects can do to accelerate things. We may be able to combine these operations into a single instruction that is faster than a software shuffle followed by a SIMD arithmetic operation (e.g., Intel's SSE3 haddps). However, the instruction is probably still going to be slower than either a data movement or SIMD arithmetic operation in isolation. While we may consider combining multiple levels of the tree reduction in one instruction, this is likely to provide no further benefit, given the required serialization of operations.

5.4   REDUCING CONTROL DIVERGENCE

In Chapter 3, we introduced control divergence, and described how it can negatively impact SIMD efficiency. We now describe how to leverage horizontal operations, as well as gather and scatter operations, to reduce control divergence.

Often, the easiest way to reduce control divergence for a given algorithm and input is to use software to reorder *all* elements being processed (i.e., not just those in each vector) so that afterwards, a given vector is likely to have more elements taking the same path through the program. If we perform some work on all data items before the point of control divergence, then we may incorporate this reordering into that operation. Otherwise, we may add reordering as a pre-processing step. Either way, this will incur instruction and execution time overhead, but may decrease the total instruction count and/or execution time. The odds of this go up if the SIMD computation is repeated multiple times on the same input; that allows us to amortize the overhead of pre-processing.

Figure 5.7 shows an example loop where we have applied this technique, although instead of reordering elements, we make a copy of the subset that pass the condition. The pre-processing step checks whether each original scalar iteration passes the condition, and if so, it stores the scalar iteration number into a buffer. After this pre-processing step, we thus know how many of the original scalar iterations we need to perform computation on, and which iterations those are. Those two pieces of information are the critical ones to allow us to perform the computation with SIMD execution with no control divergence.

In the example, we also copy into buffers the data inputs for the iterations that pass the condition. Since the scalar iterations that pass the condition are not necessarily contiguous, this gather operation guarantees that the vectorized code can perform contiguous loads from the data buffers. This gather is necessary for SIMD execution, but does not necessarily have to be integrated into the pre-processing step.

Similarly, the output data produced by the vector loop needs to be written back to potentially non-contiguous memory locations. In our example, we perform this scatter operation in line with the vector loop. That is, as soon as we produce the results, we write them back to memory—we use a scalar loop at the end of each iteration of the vectorized loop. We could instead use a similar approach as we did for the gather; we could buffer all of the results from the vectorized loop, and then have a separate loop to scatter all of the output elements. The inline approach allows us to use a smaller buffer—we need only buffer $VLEN$ elements with the inline approach, whereas a post-processing step wold require a buffer large enough to hold all of the elements that pass the condition.

While our example and much of the remaining discussion in this section centers around using software (with some hardware assistance) to regroup elements to boost efficiency of SIMD computation in the presence of control divergence, there are alternatives. In particular, there are a number of proposals for hardware to automatically detect control divergence and either skip over masked-out elements or perform regrouping automatically.

```
// Scalar loop
for (i = 0; i < N; i++)
{
    if (A[i] > 0)
    {
        B[i] += C[i];
    }
}

// Pre-processing step including data copying
for (i = 0; i < N; i++)
{
    if (A[i] > 0)
    {
        buffer_index[num_elements] = i;
        buffer_B[num_elements] = B[i];
        buffer_C[num_elements] = C[i];
        num_elements++;
    }
}

// Vectorized computation
for (i = 0; i < num_elements; i += VLEN)
{
    vload v0, buffer_B[i]
    vload v1, buffer_C[i]
    vadd v2, v0, v1
    vstore buffer_results, v2

    for (j = 0; j < VLEN; j++)
    {
        index = buffer_index[i + j];
        B[index] = buffer_results[j];
    }
}
```

Figure 5.7: Example of vectorized loop using pre-processing step to regroup elements.

Smith et al. [2000] proposed hardware support that detects power-of-two blocks of contiguous zeros in a mask register and skips over them. However, skipping over masked-out elements only reduces execution time for processors that spread execution of a vector across multiple ALUs and/or multiple cycles on the same ALU.

There has been intense recent interest in automatic regrouping, although this has so far been restricted to processors with implicit masking. Current CPU instruction set architectures and programming models require that software have control over and visibility of the contents of each vector, which seems to prohibit automatic regrouping. On the other hand, some GPUs long ago adopted the implicit vector model, and hardware to automatically handle control divergence when executing under it [Levinthal and Porter, 1984]. This allows automatic regrouping without any software side effects. As in the software approach shown above, the idea is that when an application processes many more data items than the maximum vector length, we can combine multiple vectors to fill in "holes" in the mask, and reduce the number of vectors needed for a computation. To avoid some of the complexity of the gather/scatter approach discussed above, many of the techniques regroup elements between vectors, but restrict an element to keep its position within a vector [Diamos et al., 2011, Fung et al., 2007, Fung and Aamodt, 2011]. That limits how much regrouping can be done, so more recent proposals include permuting vectors to obtain more aggressive regrouping [Rhu and Erez, 2013, Vaidya et al., 2013].

Quantifying the Benefits of Regrouping

Figure 5.8 shows the performance effects of regrouping elements. This assumes perfect regrouping, where we achieve 100% SIMD efficiency for the computation loop after regrouping. We combine the pre-processing overhead and any additional overhead, such as the scatter operation to write back results, and express the time for those steps as a percentage of the time taken for the SIMD loop (before regrouping). We show the performance with regrouping in two ways: (1) as the resulting SIMD efficiency, including the impact of the regrouping overhead—these are the lines in the graph; and (2) as speedup over the original code—these are the bars in the graph. We show these performance results for a range of control divergence, expressed as the SIMD efficiency of the original code: from 99% (blue) down to 25% (gold).

For code with low original SIMD efficiency from control divergence, we see a big performance boost from reordering, as long as the cost of reordering is not extreme. Since this technique boosts the efficiency of the compute part of the vectorized loop to 100%, if we ignore the overhead of regrouping, our speedup is $\frac{1}{original\ efficiency}$. For instance, with an original efficiency of 25%, we can obtain a 4x speedup. However, the regrouping overhead eats into this benefit. With only 8% overhead, our 4x speedup drops to about 3x.

This overhead means that, for code with little control divergence, there is little to be gained from regrouping, and plenty of performance to lose. If the original SIMD efficiency is 99%, the break-even point is 1% overhead—if the overhead is any higher, the code with regrouping is slower than the code without regrouping.

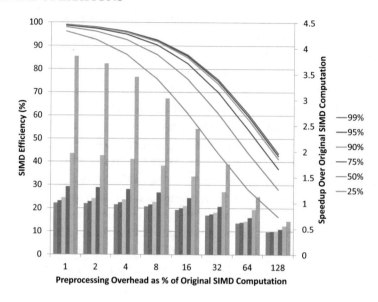

Figure 5.8: SIMD efficiency (lines) and speedup (bars) when using a pre-processing step to regroup data items.

Nested Control Flow and Regrouping

We can apply regrouping even in the presence of nested control flow. In the simple case where we only perform computation on elements that pass multiple conditions, we can simply evaluate all of the conditions in our pre-processing step. Things are more complicated when we need to perform some computation between the different condition checks.

Figure 5.9 shows an example loop with nested control flow, and where we have computation at each level of nesting. We can use regrouping on the first condition to vectorize the first computation, the line computing `temp`. However, the next condition may reintroduce control divergence by further filtering the set of input data items. We can prevent this by applying another regrouping step.

Figure 5.10 shows the contents of the arrays and buffers if we apply two stages of regrouping in our example computation. Darker cells in the figure indicate elements that pass a condition. At the first condition, we filter A into `buffer_A`, and record the indices in `buffer_index_1`. The vectorized first computation produces a set of values of `temp`. We test these in the second condition, writing the values that pass into `buffer_temp`, and the corresponding *original scalar indices* into `buffer_index_2`. The vectorized second computation produces a set of values that we finally scatter back into B, using the indices from `buffer_index_2`.

```
for (i = 0; i < N; i++)
{
    if (A[i] > 0)
    {
        temp = A[i] & bit_mask;

        if (temp > constant)
        {
            B[i] = -temp;
        }
    }
}
```

Figure 5.9: Example computation with nested control flow and computation at each level.

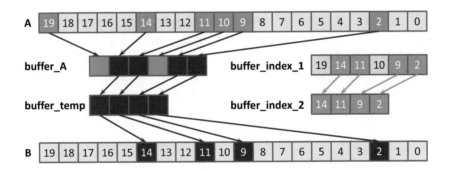

Figure 5.10: Contents of arrays and buffers in nested regrouping example. The contents are ordered from right to left.

Reducing Regrouping Overheads

Regrouping is a powerful technique to reduce or even eliminate control divergence, but it carries potentially high costs. The main overheads spring from converting to and from contiguous buffers used as inputs and outputs, respectively, of the SIMD computation loop. The buffers themselves are also an issue because they may occupy a large amount of memory, and pollute caches.

We first address the conversion process, which up to now, we've shown as a scalar operation. We may use SIMD execution for this, and thus speed it up considerably. The conversion process involves four steps.

1. Identify the data items that pass the condition.

2. Build a list of identifiers (e.g., scalar loop indices) for those data items, including counting the number of items in the list.

3. Gather the corresponding data items into contiguous memory locations, or directly into SIMD registers.

4. After the computation is done, scatter the results back to possibly non-contiguous memory locations from a contiguous buffer or directly from SIMD registers.

The first step simply requires SIMD comparisons, and the last two steps can be vectorized with gathers and scatters. That leaves the second step.

We can view the list building step as taking an array of all the loop indices and a bit per index indicating whether we'd like to perform computation on that index, discarding the undesired indices, and packing the remaining indices into a contiguous list. The inputs are easy to obtain—it is trivial to generate the list of indices in SIMD fashion, and we can generate a mask for each vector of loop indices using SIMD comparison instructions. Vectorizing the pack operation is easy with SIMD compress instructions.

Example 5.3   Figure 5.11 shows an example SIMD regrouping operation using a SIMD compress instruction, for a loop that originally tests for $A > 0$. We load a set of values from A, test them against zero to produce a mask, and compress our current loop indices, contained in v2, using the mask. We then append the compressed indices to the buffer of valid indices, and update the number of elements in the buffer by the number of bits that are in our mask (using popcnt to count the bits). Finally, we update our loop indices for use in the next iteration.

```
for (i = 0; i < N; i += VLEN)
{
    vload v3, A[i]
    vcmpgt m0, v3, v0 // v0 contains all zeros
    vcompress v4, m0, v2
    vstore buffer_index[num_elements], v4

    num_elements += popcnt(m0);

    vadd v2, v2, v1
}
```

Figure 5.11: Example SIMD regrouping operation. All elements of v0 are zero, all elements of v1 are $VLEN$, and v2 initially contains 0, ..., $VLEN - 1$.

The remaining overhead that we may want to address is the space used to hold the compressed indices. If our original array has many elements, and a large fraction of the elements pass the condition(s), then the buffer holding compressed indices may be large. If we want to construct this buffer once and reuse it many times, there are no obvious alternatives. However, if we do not plan on reusing the compressed indices, or if we can construct them cheaply enough that the space overhead is a bigger concern, then we may make do with a small buffer.

We need a buffer containing only twice as many elements as our vector length if we consume a vector's worth of elements as soon as we produce them [Plotnikov et al., 2014]. That is, we can use this algorithm to perform regrouping in line with the SIMD computation.

1. Test $VLEN$ elements and append the indices of the passing ones to the buffer.

2. If the buffer has at least $VLEN$ elements, or if we've tested all elements, perform one iteration of SIMD computation.

If we use this approach, our buffer can be quite small. We cannot know *a priori* how many elements will pass each SIMD condition check; thus, we cannot know how many elements we will append to the index buffer—it may be zero, the vector length, or anything in between. However, we know that as soon as we have at least enough elements to fill a SIMD register, we will consume that many elements. Thus, we can never have more than two times the vector length minus one number of elements in the buffer.

This inlining approach works not just for the buffer of indices, but for any input or output buffers we elect to use. We say "elect to use" because we may not need to build input or output buffers besides the index buffer. With gather and scatter instructions, we may choose to skip building buffers for input and output values. Instead, we can use our buffered indices to gather our inputs from memory into SIMD registers, and once we've done our SIMD computation, scatter the results back to memory from SIMD registers.

5.5 POTENTIAL DEPENDENCES

Horizontal SIMD data movement and reduction instructions allow software to satisfy *known* dependences between elements in a vector, including when software must compute at run-time *which* element is dependent upon which. However, some computations have *unpredictable* dependences. That is, at compile-time, the software does not know whether a dependence exists at all, only that there *might* be a dependence or dependences. The most common source of this situation is indirect memory operations where the indirect pattern is not known statically. These introduce *potential dependences* in a computation, which can prevent vectorization.

Figure 5.12 shows a histogram computation, a canonical example of this pattern. For each element of the input array, we map it to a bin in the histogram (in this case, the mapping is the identity function), load that bin, increment it, and write back the updated value. A naive programmer might vectorize this as shown at the bottom of the figure. We operate on a vector of input elements at a time, reading them in, mapping them to bins, gathering those bins, incrementing them, and scattering back the updated values. Figure 5.13 shows an example iteration of this approach, along with the correct result (i.e., from scalar execution) for those elements.

```
// Scalar loop
for (i = 0; i < N; i++)
{
    bin = A[i];
    histogram[bin]++;
}

// Incorrectly vectorized loop
for (i = 0; i < N; i+=VLEN)
{
    vload v0, A[i]
    vgather v1, histogram, v0
    vadd v1, v1, v2  // v2 is {1, 1, ..., 1}
    vscatter histogram, v0, v1
}
```

Figure 5.12: Example histogram computation.

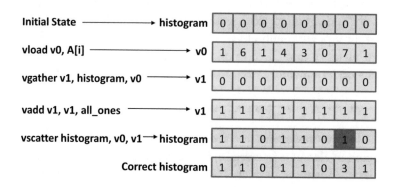

Figure 5.13: Example execution of the incorrectly vectorized histogram.

The source of the error is that the original scalar loop contains possible read-after-write (RAW) dependences between the write to the histogram in one iteration and the reads from it in the following iterations. A *possible* dependence becomes a *real* dependence when a set of inputs in the same vector contains any values that map to the same bin. In that case, we perform the read-update-write (a.k.a. read-modify-write) to the histogram for the same bin multiple times simultaneously, thereby reading it twice, doing the updates, *then* writing it twice, when we should have performed two distinct read-update-write pairs in serialized fashion.

To beneficially vectorize computations like this, we must be able to efficiently dynamically detect and enforce any real dependences, or *conflicts* between vector elements. The cost in instructions for each vector iteration (i.e., each VLEN scalar iterations) is *conflict detection insts* $+ \frac{original\ insts}{SIMD\ efficiency} + conflict\ handling\ insts$, where the denominator of the middle term is the SIMD efficiency of the computation absent the conflict detection and enforcement. Thus, especially for small loops, the conflict detection and handling costs are critical to achieving good SIMD efficiency.

Our discussion of conflict detection and handling will focus primarily on software techniques, with some hardware acceleration. This is similar to the approach taken in Kumar et al. [2008]. That work combined gathers with load-linked instructions, and scatters with store-conditional instructions, to allow each element's read-modify-write operation within a vectorized loop to happen in an atomic manner, ensuring no dependences could be violated. Similarly, hardware could provide a single primitive to perform atomic read-update-write operations in SIMD fashion. Ahn et al. [2005] proposed this, called *scatter-add*, along with an implementation involving functional units in cache and/or memory controllers.

Alternatively, we could use a pure hardware approach. Pajuelo et al. [2002] proposed hardware to speculatively vectorize loops containing strided loads and stores (i.e., no gathers or scatters). The hardware then checks for dependence violations by using a content addressable memory (CAM), similar to a load or store buffer, to detect illegal reordering of loads and stores to the same address.

5.5.1 SINGLE-INDEX CASE

We first consider loops similar to our histogram example, where inside the loop, we perform a simple read-update-write. That is, in the scalar code, we read a memory location, perform a computation on that value, and write the result back to the original memory location. When vectorized, this becomes a gather-update-scatter, where the gather and scatter share the same base address and index vector.

In this case, the only dependence type we need to detect is RAW, as in our earlier example. These dependences manifest only when we attempt to perform more than one read-update-write operation with the same index using SIMD execution. Thus, we need only detect when we have duplicate, or conflicting, indices in the same index vector.

Our choice of detection technique will depend on how we plan to enforce any dependences we find. For example, if we plan to fall back to scalar execution if we find *any* dependences, then we need only determine whether any duplicate indices exist in a given index vector. However, if we plan to use SIMD execution in the presence of dependences, to exploit as much parallelism as possible, we need more information about any conflicts.

Conflict Behavior

Before we explore techniques for detecting and enforcing dependences, we look at expected conflict behavior. This gives us some idea of how helpful vectorization can be in the presence of potential dependences.

Let us assume that our loop updates all elements in an array with uniform probability. One useful measure is the expected fraction of index vectors containing any duplicates. If we use a different execution path in the case of conflicts, this tells us what fraction of vectors will use that fallback path. Another useful measure is the expected frequency of the most common index in each vector. That is, for a given vector, find the average number of occurrences of the index that appears the most times. Our parallelization of each vector is limited by this number. If this number is one, then we expect all indices to be unique. If this number is the vector length, then we expect each index vector to comprise only a single value.

Figure 5.14 shows these two measures, for several different array sizes (in elements) and vector lengths. For very small vectors, both measures are close to ideal for even very small arrays. However, for those small arrays, both measures grow very fast with vector length. For example, for an array with 16 elements and a vector length of 32, the expected number of occurrences of each index is only two, but the expected *maximum* frequency is almost five, much worse news for vectorization. The behavior is much better for larger arrays: at 4096 elements, the fraction of vectors with any conflicts is less than 3% up to a vector length of 16, and the expected maximum frequency is only 1.1 for a vector length of 32.

These results are promising for vectorization of loops with possible dependences through indirect memory accesses. However, the bad news is that in real applications and data sets, we often have non-uniform index distributions. Even for very large arrays, if indices comes in bursts of similar values, we are likely to suffer significant conflicts. Thus, it pays to consider the best approaches for both detecting and handling conflicts.

Detecting Conflicts

The most straightforward way to detect duplicate indices is with a brute force scalar comparison loop. Figure 5.15 shows example code to search a single index vector for conflicts. For each index, we check for equality with *earlier* indices in the vector. This prevents redundant comparisons—since we only want to know if any indices are duplicates, if we compare index i against index j, there's no reason to later on compare j to i. There's also no reason to compare an index against itself. Thus, the total number of comparisons is $\frac{VLEN \times (VLEN-1)}{2}$. While the code shown only detects whether any conflicts exist, we may easily modify it to determine and record which indices are duplicates of which.

We may use SIMD execution to accelerate this brute force approach. Figure 5.16 shows example code implementing a method to broadcast each index in turn and compare it against all others, checking for any duplicates. We must discard the comparison result of the index against itself, but otherwise, any matches we find indicate a conflict. This technique uses $VLEN$ compari-

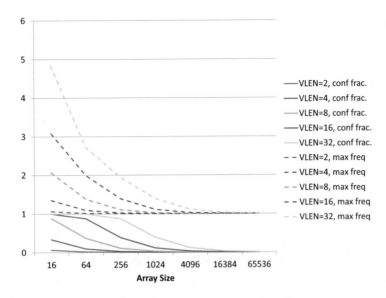

Figure 5.14: Fraction of SIMD iterations with at least one conflict (solid lines) and average frequency of most common index in a vector (dashed lines).

```
for (i = 0; i < VLEN; i++)
{
    for (j = 0; j < i; j++)
    {
        if (index[i] == index[j])
            return true;
    }
}
```

Figure 5.15: Scalar code to detect conflicts.

son instructions to perform $VLEN^2$ comparisons; unlike the scalar loop, this performs redundant comparisons since we compare all indices to all others, but avoiding the redundant comparisons is more difficult than doing them. Like the scalar version, this method can be easily modified to count duplicates, or find and record which indices are duplicates of which.

To further reduce the instruction cost of this method of conflict detection, we can add a SIMD instruction to perform all of the needed comparisons (e.g., the Intel AVX-512 vpconflict instruction; Intel [2014c]). To implement this, we either need $\frac{VLEN \times (VLEN-1)}{2}$ comparators, or we need to occupy a smaller set of comparators for multiple cycles and permute the inputs

```
// v0 = indices to test
for (i = 0; i < VLEN; i++)
{
    vbroadcast v1, v0, i
    vcmpeq m0, v0, v1

    if ((m0 & ~(1 << i) != 0) // check for match besides self-match
        return true;
}
```

Figure 5.16: SIMD version of conflict detection using a broadcast and compare approach.

appropriately. Depending on what form the output of this instruction takes, we may be able to process it to count duplicates or find which indices are duplicates of which.

Another conflict detection method is to make use of one of the very properties we're worried about: a scatter with duplicate indices will have one element overwrite others. Figure 5.17 shows example code implementing a method which performs a scatter and then a gather to detect any overwrites. The idea is to scatter a vector of unique values to a buffer using the indices under

```
// v0 = {0, 1, ..., VLEN-1}
// v1 = indices to test
vscatter buffer, v1, v0
vgather v2, buffer, v1
vcmpeq m0, v2, v0
if (m0 != all_ones)
    return true;
```

Figure 5.17: Gather/scatter approach to detect conflicts.

test—if we have any duplicate indices, the subsequent gather will read the same value multiple times, and the result will not match the original vector [Smelyanskiy et al., 2009b]. Figure 5.18 shows an example execution of this method. With this approach, we need only a single SIMD comparison instruction, but we also need a scatter and a gather instruction, which are relatively expensive. This method also requires a memory buffer, which could be quite large. We can reduce the storage requirements by limiting the indices to a certain number of bits, e.g., a 16 element buffer is sufficient if we only use the least significant four bits of each index. This, however, discards information and can lead to false conflicts. Rather than just providing a boolean conflict/no conflict answer, this method can be used to provide us a mask indicating a maximal set of unique indices (i.e., m0 in the example). For duplicate indices, the gather results also indicate which index

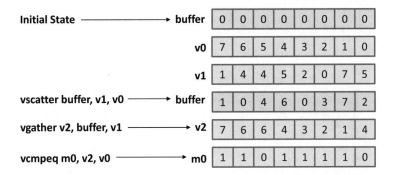

Figure 5.18: Example of the gather/scatter approach to detect conflicts.

they match. However, this method is more difficult than the others to modify to obtain a count of duplicates.

Scalar Execution

To guarantee correctness in the presence of conflicts, we may choose to use scalar execution. For a vectorized loop where we detect a conflict in a given vector, we may fall back to scalar execution for just that vector, for that vector and all future iterations of the loop, or anywhere in between. Regardless, this techniques is easy to employ—we simply check whether we have a conflict or not using one of the above techniques and jump to the fallback execution path.

Example 5.4   Figure 5.19 shows the SIMD efficiency when we fall back to scalar execution for any vector with any conflicts, and then resume SIMD execution for the next vector, for various conflict frequencies. This assumes a vector length of 16, a cost of detecting conflicts of either 32 instructions (solid lines) or one instruction (dashed lines), and that the original scalar loop can otherwise be vectorized at 100% efficiency. With a high cost of conflict detection, SIMD efficiency is low, regardless of conflict frequency, unless the amount of computation in the loop is large. Even then, the use of scalar execution for the fallback path limits SIMD efficiency to $\frac{1}{conflict\ ratio \times VLEN}$. For example, with 50% of vectors having conflicts, we will never exceed 12.5% efficiency (i.e., one eighth) because we can never reduce the instruction count by more than a factor of two, and the vector length is 16.

Since a scalar fallback has such a dramatic effect on SIMD efficiency in the presence of a significant number of conflicts, we may choose to use scalar execution only if we detect *enough* duplicates. This could mean detecting either enough index elements that are not unique, or that the most common index in a vector has enough copies. The latter is generally a more useful approach, as we will discuss later. Conditionally using the scalar fallback when we exceed a threshold

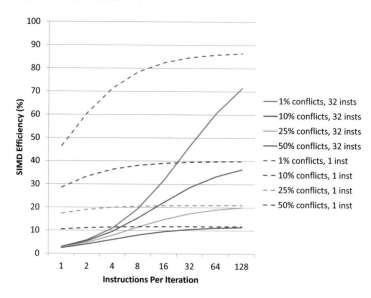

Figure 5.19: SIMD efficiency of scalar fallback approach, for various conflict frequencies.

of duplicates means we must count duplicates, and also that we must use the SIMD execution techniques presented next when we have some duplicates, but not too many.

SIMD Execution

To correctly use SIMD execution in the presence of conflicts, we must carefully choose which elements to operate on simultaneously. The most straightforward approach is to force serialization of any computation corresponding to duplicate indices—this way, we serialize any read-update-write operations to the same memory location, satisfying any RAW dependences. We can do this via the following procedure.

1. Initialize the set of vector elements to operate on, V, to everything in the vector.

2. From the elements in V, find a maximal set of indices with no duplicates.

3. Use SIMD execution for the data elements corresponding to the set found in step 2 (e.g., with masking).

4. Remove the elements just operated on from V, and if V is non-empty, go back to step 2.

This adds a loop inside the vectorized loop, to serialize execution of elements with duplicate indices. This innermost loop uses SIMD execution on a set of elements with unique indices, guaranteeing correctness.

Example 5.5 Figure 5.20 shows an example execution of this approach when used for a histogram computation. The right-hand side shows the histogram in memory after each iteration. The left-hand side shows the index vector (i.e., bin), with light shaded elements still to be operated on, and dark elements completed. The index vector includes three copies of the value 1; all other indices are unique. The first iteration operates on all indices except for two of the 1s. The second iteration operates on one of the remaining 1s, and the final iteration operates on the remaining one. The result matches what we get from scalar execution (see Figure 5.13).

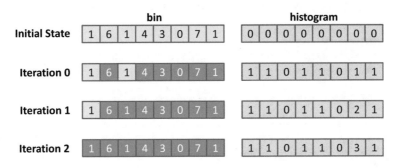

Figure 5.20: Using SIMD execution in the presence of conflicts. Computation on duplicate indices is serialized.

As shown in the example, this procedure may reorder the execution of elements relative to the original scalar order. When we use SIMD execution, we already have permission (e.g., from the programmer) to perform computation simultaneously that is done in sequence with scalar execution. However, the architectural state never reflects a *reordering* of scalar iterations, only that a set of them are executing simultaneously. This procedure may actually reorder scalar iterations, and debuggers, fault handlers, and other software that may view program state in the middle of execution must be aware of this if they view a snapshot of architectural state in the middle of the procedure.

Data communication satisfying a RAW dependence from one vector element to another may be done through either memory or registers. To communicate via memory, the SIMD execution can use gathers and scatters to turn the scalar read-update-write operations into gather-update-scatter operations. With an appropriate mask, we will read and write a unique set of elements with each gather and scatter. An element that is forced to serialize will have its gather read a just-scattered value from an element with the same index, satisfying the RAW dependence.

To communicate via register, we need to do a little more work on the conflict detection side, but we can avoid extra reads and writes to memory. For each vector of data elements, we use a single gather to read them from memory. The very first iteration of the procedure uses these values for its input. However, later iterations are all operating on elements with duplicate indices—their inputs can be read from the output register(s) of the previous iteration. As part of the step of finding a maximal set of unique indices (from those remaining), for each element, we determine with which element it previously conflicted. We use a permute to move the output(s) from the conflicting element into the correct position in the input vector(s) to satisfy the RAW dependence. Finally, at the end of the procedure, we perform a single scatter to write back the latest results for each unique index.

Example 5.6   Figure 5.21 shows the SIMD efficiency when we use SIMD execution for all elements, for various average numbers of SIMD iterations (i.e., the innermost loop) to complete each vector. The figure assumes a vector length of 16, a cost of detecting conflicts of either 32

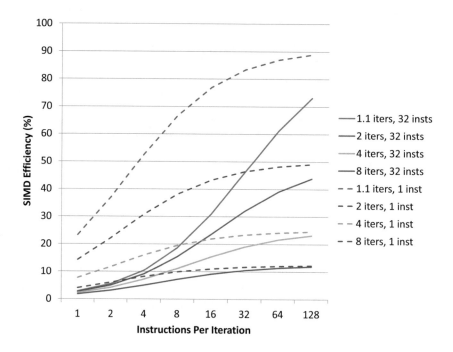

Figure 5.21: SIMD efficiency when using SIMD execution even in the presence of conflicts, for various conflict frequencies.

instructions (solid lines) or one instruction (dashed lines), that each SIMD iteration includes two overhead instructions for bookkeeping (i.e., finding the next set of indices and updating the set of remaining indices), and that the original scalar loop can otherwise be vectorized at 100% efficiency. These results are similar to the previous example showing SIMD efficiency when using scalar execution when there are conflicts. However, the number of inner loop iterations here is *not* the same as the conflict rate that parameterizes the scalar fallback. In the scalar fallback approach, even a single duplicate may force us to abandon SIMD execution; here, even if *every* vector has some duplicates, the average number of iterations of the innermost loop may be only two. For example, if an index vector contains two of every index value, we need only two SIMD iterations to compute all elements in the vector. As the number of instructions in each original scalar iteration rises, we amortize the conflict detection and bookkeeping overheads, and SIMD efficiency approaches $\frac{1}{innermost\ loop\ iterations}$.

If conflicts are frequent, but are not concentrated on a small set of indices, the serialization approach may allow for reasonable SIMD efficiency. However, if conflicts are concentrated, then serialization has worst-case behavior that's no better than scalar execution. If we have an index vector with only one value, we need as many SIMD iterations as scalar ones, and must pay the conflict detection and bookkeeping overheads, leading to significantly worse performance than if we used scalar execution.

One solution to this is to use a tree reduction approach. This is a variant of the tree reduction described in Section 5.3. Instead of combining all elements in a register, we only want to combine elements that have the same index value. Unlike the serialization approach, this *must* be done in-register. We can do this with the following procedure.

1. Initialize the set of vector elements to operate on, V, to everything in the vector.

2. For the elements in V, pair up duplicate indices; record and remove unique indices (these are the "roots" of the reduction trees).

3. Use SIMD execution to combine the elements in each pair (the same as the in-register communication approach above).

4. Remove one element from each pair from V, and if V is non-empty, go back to step 2.

5. Perform the gather-update-scatter, using only the root elements.

Example 5.7   Figure 5.22 shows an example execution of this approach. We want to add the values on the right-hand side to the memory locations indicated by the indices on the left-hand side. As we combine pairs of elements in each SIMD iteration, we indicate the removed elements by erasing them from the value vector and showing them as dark in the index vector. The index vector contains one unique index—a seven—and two with duplicates—two fours, and five ones. In the first iteration, we combine the values of the two fours, and also combine two pairs of ones, leaving only ones as duplicates. In the second iteration, we combine two of the ones, and in the final iteration, we combine the remaining two ones. We are left with three values, one corresponding to each unique index. Although not shown, we then do a gather-add-scatter to update memory.

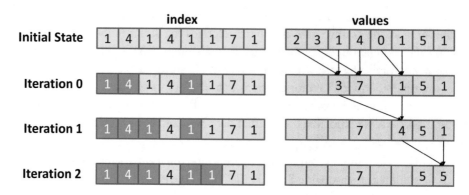

Figure 5.22: Using a tree reduction to combine values corresponding to duplicate indices.

With this approach, we do $\lceil \log_2 (frequency\ of\ most\ common\ index) \rceil$ SIMD iterations. This gives us worst case behavior of only $\log_2 VLEN$ iterations, instead of the VLEN potentially needed for the serialization approach. The tradeoff is additional overhead for pairing up duplicate indices.

5.5.2 MULTI-INDEX CASE

Conflict detection becomes more complicated if we use multiple indices for the same array in the scalar loop. Multi-index dependences are harder to detect because the computation and data movement can be more complex than gather-update-scatter; within a scalar iteration we may read and write different elements of the same array. Figure 5.23 shows an example of such a computation, where we simply copy data elements in array A from indices in C to indices in B. Unless we can guarantee that within a vector, we have no duplicate values between B and C, we may have a data dependence and the naive vectorization at the bottom of the example may give a wrong answer.

Example 5.8 Figure 5.24 shows an example execution of the (incorrect) SIMD code next to the correct scalar execution. The top row shows the initial state of A, as well as the states of B and C, which do not change. The rows below show the updated state of A. For each iteration, the darkly shaded elements are the ones just written. The SIMD code produces one incorrect element, pointed to with a red arrow. This error arises because the second write to A[1] (i.e., iteration 2 in scalar execution) is dependent on the write to A[5] (i.e., from iteration 1 in scalar execution), and we violate that dependence. The rightmost column shows a correct SIMD execution for this vector. The first SIMD iteration only operates on the first two data elements and stops there to honor the dependence; the second iteration operates on the remaining elements.

Like the single index case, conflicts derive from duplicate indices; however, in this case, the duplicates may be present in different vectors. Duplicate indices in the *same* element of different

```
// Scalar loop
for (i = 0; i < N; i++)
{
    A[B[i]] = A[C[i]];
}

// Incorrectly vectorized loop
for (i = 0; i < N; i+=VLEN)
{
    vload v0, B[i]
    vload v1, C[i]
    vgather v2, A, v1
    vscatter A, v0, v2
}
```

Figure 5.23: Example multi-index computation.

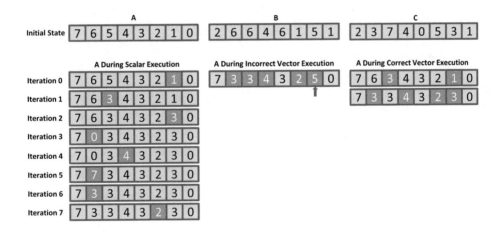

Figure 5.24: Example execution of the scalar and SIMD loops with multiple indices, and correct SIMD execution.

index vectors is not a problem—those reads and writes will happen in the original scalar order. Instead, we are concerned about duplicates in different positions in different index vectors.

To detect conflicts, we can check all index vectors against all others, and each vector against itself, for duplicate indices (excepting elements in like positions); however, this is overkill. We need only detect if a reordered read or write violates a dependence. Thus, we need only check each index against *earlier* elements in index vectors used *later* in the loop—when vectorized, we reverse the order of use of these indices. For example, in our code example, we need only check each index from C against earlier elements of B. Since we read A[C[i]], if an earlier (scalar) write to A[B[i]] is to the same location, we should read that result, but won't in the vectorized loop where the gather happens before the scatter.

For a loop with N index vectors, this requires $\frac{N\times(N-1)}{2}$ conflict checks, each of which is as expensive as a single index conflict check. Those checks can use similar instruction support as in the single index case, but must take two input vectors [Hughes et al., 2012].

Things are more complicated if we try to wring all possible parallelism out of a vector. Recall that in the single index vector case, we are free to reorder the original scalar iterations (i.e., choose an arbitrary set of elements to operate on in any vector iteration). Reordering can sometimes allow us to exploit more parallelism, e.g., if the index vector was x, x, y, y, then unless we operate on one "x" and one "y" simultaneously, we will take at least three SIMD iterations instead of the minimum two. However, when we reorder scalar iterations we reorder more loads and stores than we described above—we no longer only execute accesses farther down the loop body early, but may execute accesses farther up the loop body late. In the multi-index case, this means we may need to check all elements in all index vectors against each other. For example, returning to Figure 5.24, if we reorder the execution of the sixth and seventh elements (both with B[i] of 6), then A[6] will have a wrong value after we complete execution, seven instead of three, because we will have violated a write-after-write, or WAW, dependence.

CHAPTER 6

Conclusions

Architecture evolution is driven by the changing needs of users, and thus, changing application characteristics. In recent years, SIMD has become a critical part of the performance and energy efficiency picture for general-purpose processors, especially in high-performance computing. This was triggered by the importance of algorithms involving plenty of very regular data parallelism, such as operations on long streams of media data and large dense matrices. Further, the effectiveness of SIMD on such algorithms has motivated longer vectors. This is despite long vectors' lower efficiency for codes with limited or more complex data parallelism, such as codes operating on sparse matrices/vectors and graphs. To improve the performance and efficiency of SIMD for these codes, architects have finally introduced to commodity hardware support for non-contiguous memory operations and horizontal operations.

6.1 FUTURE DIRECTIONS

The future of SIMD in general-purpose processors depends on which applications are seen as most critical. For example, much recent attention has turned to Big Data analytics, including applications employing machine learning or artificial neural networks. These applications have plenty of data parallelism, and are good candidates for SIMD execution, but the most effective design varies across applications. Some of these applications, such as image classification via a convolutional neural network, are dominated by regular data parallelism. Others, such as summarizing properties of social networks represented as very sparse graphs or matrices, have copious irregular parallelism.

We believe both regular and irregular data parallelism are likely to remain too critical to ignore for the foreseeable future. Many problems map nicely to regular parallel algorithms, and with traditional SIMD hardware, we can execute them with high efficiency. Thus, abandoning efficient support for regular memory operations and longer vectors would likely hurt many users. At the same time, there are always problems that map very poorly to regular parallel hardware, and will benefit from better support for irregular parallelism.

While there are certain properties of SIMD architectures that must be selected for the expected available parallelism in the target applications—the key one being SIMD width—plenty of features do not involve a tug-of-war between regular and irregular data parallel applications. Certainly, hardware devoted to features for one type of application consumes area and power budget, not to mention design resources, and thus those are not available for features targeting another type of application. However, many powerful SIMD features don't require gargantuan hardware. Further, as transistor counts continue scaling, we are likely to have area for more specialized features. Thus, we expect that future designs will continue incorporating features to improve performance or efficiency for some applications without sacrificing it for others. However, where designers *must* make a call to optimize for a particular point in design space, we expect the pendulum to swing back and forth, with the relative importance of different application domains.

In optimizing future SIMD architectures for regular parallelism, we would expect to see a push for increased compute density. In addition to longer vectors, this may motivate improved support for smaller data types. If we can pack the needed data into fewer bits, we can perform more operations simultaneously with the same SIMD width. We could also see SIMD instructions that pack more computation into one instruction, in a CISC (complex instruction set computing) sense. This could allow for higher compute density without requiring longer vectors or wider issue widths, mitigating the negative impact to applications with limited data parallelism.

In optimizing future SIMD architectures for irregular parallelism, we would expect to see a push for improved efficiency under non-ideal circumstances, such as non-contiguous memory operations and control divergence. While we might expect shorter vectors, we could instead see just narrower SIMD execution hardware. If we optimize for lots of partly filled vectors due to limited parallelism, we can maintain instruction throughput at higher efficiency with smartly controlled narrow execution units. For example, we may automatically compress and expand the elements in each vector, to require as few passes as possible through narrow execution hardware. We also expect to see better gather and scatter support, including specialized hardware and instructions for various locality scenarios like AoS or strided accesses, and also caches or software buffers designed to support more simultaneous accesses. Finally, we expect improved support for horizontal operations that enable vectorization where it's not easy or profitable today.

Bibliography

J. H. Ahn, M. Erez, and W. J. Dally, "Scatter-Add in Data Parallel Architectures," *Proceedings of the 11th International Symposium on High-Performance Computer Architecture*, 2005. DOI: 10.1109/HPCA.2005.30. 81

J. R. Allen, K. Kennedy, C. Porterfield, and J. Warren, "Conversion of Control Dependence to Data Dependence," *Proceedings of the 10th ACM SIGACT-SIGPLAN Symposium on Principles of Programming Languages*, 1983. DOI: 10.1145/567067.567085. 37

T. M. Austin and G. S. Sohi, "High-Bandwidth Address Translation for Multiple-Issue Processors," *Proceedings of the 23rd Annual International Symposium on Computer Architecture*, 1996. DOI: 10.1145/232974.232990. 17

U. Banerjee, "Data Dependence in Ordinary Programs," MS Thesis, Report 76-837, University of Illinois at Urbana-Champaign, Department of Computer Science, November 1976. 25

G. H. Barnes, R. M. Brown, M. Kato, D. J. Kuck, D. L. Slotnick, and R. A. Stokes, "The Illiac IV Computer," *IEEE Transactions on Computers*, vol. 17, no. 8, pp. 746–757, 1968. DOI: 10.1109/TC.1968.229158. 10

C. Batten, R. Krashinsky, S. Gerding, and K. Asanovic, "Cache Refill/Access Decoupling for Vector Machines," *Proceedings of the 37th Annual IEEE/ACM International Symposium on Microarchitecture*, 2004. DOI: 10.1109/MICRO.2004.9. 66

A. Bik, M. Girkar, P. M. Grey, and X. Tian, "Automatic Intra-Register Vectorization for the Intel Architecture," *International Journal of Parallel Programming*, vol. 30, no. 2, pp. 65–98, 2002. DOI: 10.1023/A:1014230429447. 25, 26, 45

P. Bulic and V. Gustin, "On Dependence Analysis for SIMD Enhanced Processors," *Proceedings of the 6th international conference on High Performance Computing for Computational Science*, 2004. DOI: 10.1007/11403937_40. 26

B. L. Chamberlain, "A Brief Overview of Chapel," `http://chapel.cray.com/papers/Brief OverviewChapel.pdf`, 2013. 24

T. P. Chen, D. Budnikov, C. J. Hughes, and Y.-K. Chen, "Computer Vision on Multi-Core Processors: Articulated Body Tracking," *Proceedings of the 2007 International Conference on Multimedia and Expo*, 2007. DOI: 10.1109/ICME.2007.4285037. 4

Y.-K. Chen, J. Chhugani, P. Dubey, C. J. Hughes, D. Kim, S. Kumar, V. W. Lee, A. D. Nguyen, and M. Smelyanskiy, "Convergence of Recognition, Mining, and Synthesis Workloads and Its Implications," *Proceedings of the IEEE*, vol. 96, no. 5, pp. 790–807, 2008. DOI: 10.1109/JPROC.2008.917729. 1

J. Chong, K. You, Y. Yi, E. Gonina, C. Hughes, W. Sung, and K. Keutzer, "Scalable HMM-Based Inference Engine In Large Vocabulary Continuous Speech Recognition," *Proceedings of the 2009 International Conference on Multimedia and Expo*, 2009. DOI: 10.1109/ICME.2009.5202871. 5

M. A. Cornea-Hasegan, R. A. Golliver, and P. Markstein, "Correctness Proofs Outline for Newton-Raphson Based Floating-Point Divide and Square Root Algorithms," *Proceedings of the 14th IEEE Symposium on Computer Arithmetic*, 1999. DOI: 10.1109/ARITH.1999.762834. 30

Cray Research, Inc. "The Cray-1 Computer System," *Publication No. 2240008B*, 1976. DOI: 10.1145/359327.359336. 34, 36, 37

Cray Research, Inc., "CRAY-2 Computer Systems Functional Description Manual," *Publication HR-02000-0D*, 1989. 48

Cray Research, Inc. "Cray Fortran Compiler," 1977. 23

K. Czechowski, V. W. Lee, E. Grochowski, R. Ronen, R. Singhal, R. Vuduc, and P. Dubey, "Improving the Energy Efficiency of Big Cores," *Proceedings of the 41st Annual International Symposium on Computer Architecture*, 2014. DOI: 10.1145/2678373.2665743. 11

G. Diamos, B. Ashbaugh, S. Maiyuran, A. Kerr, H. Wu, and S. Yalamanchili, "SIMD Re-Convergence at Thread Frontiers," *Proceedings of the 44th Annual IEEE/ACM International Symposium on Microarchitecture*, 2011. DOI: 10.1145/2155620.2155676. 75

K. Diefendorff, P. K. Dubey, R. Hochsprung, and H. Scales, "AltiVec Extension to PowerPC Accelerates Media Processing," *IEEE Micro*, vol. 20, no. 2, pp. 85–95, 2000. DOI: 10.1109/40.848475. 70

A. E. Eichenberger, P. Wu, and K. O'Brien, "Vectorization for SIMD Architectures with Alignment Constraints," *Proceedings of the ACM SIGPLAN 2004 Conference on Programming Language Design and Implementation*, 2004. DOI: 10.1145/996841.996853. 25, 45

C. Eoyang, R. H. Mendez, and O. M. Lubeck, "The Birth of the Second Generation: The Hitachi S-820/80," *Proceedings of the 1988 ACM/IEEE Conference on Supercomputing*, 1988. DOI: 10.1109/SUPERC.1988.44666. 37

R. Espasa, F. Ardanaz, J. Emer, S. Felix, J. Gago, R. Gramunt, I. Hernandez, T. Juan, G. Lowney, M. Mattina, and A. Seznec, "Tarantula: A Vector Extension to the Alpha Architecture," *Proceedings of the 29th Annual International Symposium on Computer Architecture*, 2002. DOI: 10.1109/ISCA.2002.1003586. 17, 60, 63

A. T. Forsyth, B. J. Hickmann, J. C. Hall, and C. J. Hughes, "Coalescing Adjacent Gather/Scatter Operations," United States Patent Application no. 13/997784, filed 2012, published 2014. 66

N. Foster and D. Metaxas, "Realistic Animation of Liquids," *Graphical Models and Image Processing*, vol. 58, no. 5, pp. 471–483, 1996. DOI: 10.1006/gmip.1996.0039. 2

Free Software Foundation, Inc., `https://gcc.gnu.org/onlinedocs/gcc-4.9.2/gcc/Targ et-Builtins.html`. 23

W. W. L. Fung, I. Sham, G. Yuan, and T. M. Aamodt, "Dynamic Warp Formation and Scheduling for Efficient GPU Control Flow," *Proceedings of the 40th Annual IEEE/ACM International Symposium on Microarchitecture*, 2007. DOI: 10.1109/MICRO.2007.30. 38, 75

W. W. L. Fung and T. M. Aamodt, "Thread Block Compaction for Efficient SIMT Control Flow" *Proceedings of the 17th International Symposium on High Performance Computer Architecture*, 2011. DOI: 10.1109/HPCA.2011.5749714. 75

A. Gandhi, H. Akkary, R. Rajwar, S. T. Srinivasan, and K. Lai, "Scalable Load and Store Processing in Latency Tolerant Processors," *Proceedings of the 32nd Annual International Symposium on Computer Architecture*, 2005. DOI: 10.1145/1080695.1070007. 52

C. Gou and G. N. Gaydadjiev, "Elastic Pipeline: Addressing GPU On-Chip Shared Memory Bank Conflicts," *Proceedings of the 8th International Conference on Computing Frontiers*, 2011. DOI: 10.1145/2016604.2016608. 62

V. Govindaraju, T. Nowatzki, and K. Sankaralingam, "Breaking SIMD Shackles with an Exposed Flexible Microarchitecture and the Access Execute PDG," *Proceedings of the 22nd International Conference on Parallel Architectures and Compilation Techniques*, 2013. DOI: 10.1109/PACT.2013.6618830. 21

R. v. Hanxleden and K. Kennedy, "Relaxing SIMD Control Flow Constraints Using Loop Transformations," *Proceedings of the ACM SIGPLAN 1992 Conference on Programming Language Design and Implementation*, 1992. DOI: 10.1145/143095.143133. 25, 34

R. G. Hintz and D. P. Tate, "Control Data STAR-100 Processor Design," *Proceedings of COMPCON*, 1972. 10

S. Hiranandani, K. Kennedy, C. Koelbel, U. Kremer, and C.-W. Tseng, "An Overview of the Fortran D Programming System," In U. Banerjee, D. Gelernter, A. Nicolau, and D. Padua, editors, Language and Compilers for Parallel Computing, Lecture Notes in Computer Science no. 589, pp. 18-34. Springer-Verlag, Berlin, August 1992. DOI: 10.1007/BFb0038655. 22

Innovative Computing Laboratory, University of Tennessee, "HPC Challenge Benchmark," http://icl.cs.utk.edu/hpcc. 22

C. J. Hughes, M. J. Charney, Y.-K. Chen, J. Corbal, A. T. Forsyth, M. B. Girkar, J. C. Hall, H. Saito, R. Valentine, and J. Wiedemeier, "Vector Conflict Instructions," United States Patent Application no. 12/976616, filed 2010, published 2012. 92

C. J. Hughes, Y.-K. Chen, M. Bomb, J. W. Brandt, M. J. Buxton, M. J. Charney, S. Chennupaty, J. Corbal, M. G. Dixon, M. B. Girkar, J. C. Hall, H. Saito, P. Lachner, G. Neiger, C. J. Newburn, R. S. Parthasarathy, B. L. Toll, R. Valentine, and J. Wiedemeier, "Gathering and Scattering Multiple Data Elements," United States Patent no. 8447962, 2013. 50, 54

Intel Corporation, "Intel Intrinsics Guide," https://software.intel.com/sites/landingpage/IntrinsicsGuide/. 23

Intel Corporation, "User and Reference Guide for the Intel C++ Compiler 14.0," https://software.intel.com/en-us/compiler_14.0_ug_c. 23

Intel Corporation, "Intel Architecture Instruction Set Extensions Programming Reference," October 2014. 38, 40, 48, 71, 83

Intel Corporation, "Intel Xeon Phi Coprocessor Instruction Set Architecture Reference Manual," September 2012. 48

Japan Agency for Marine-Earth Science and Technology, Earth Simulator Center, http://www.jamstec.go.jp/es/en/system/hardware.html. 22

T. Juan, J. J. Navarro, and O. Temam, "Data Caches for Superscalar Processors," *Proceedings of the 11th International Conference on Supercomputing*, 1997. DOI: 10.1145/263580.263595. 17

R. Keryell and N. Paris, "Activity Counter: New Optimization for the Dynamic Scheduling of SIMD Control Flow," *Proceedings of the 1993 International Conference on Parallel Processing*, 1993. DOI: 10.1109/ICPP.1993.36. 38

B. K. Khailany, T. Williams, J. Lin, E. P. Long, M. Rygh, C. W. Tovey, and W. J. Dally, "A Programmable 5012 GOPS Stream Processor for Signal, Image, and Video Processing," *IEEE Journal of Solid-State Circuits*, vol. 43, no. 1, pp. 202–213, 2008. DOI: 10.1109/JSSC.2007.909331. 22

Khronos Group, "OpenCL 2.0 Manual," https://www.khronos.org/registry/cl/sdk/2.0/docs/man/xhtml. 24, 60

C. Kim, J. Chhugani, N. Satish, E. Sedlar, A. D. Nguyen, T. Kaldeway, V. W. Lee, S. A. Brandt, and P. Dubey, "FAST: Fast Architecture Sensitive Tree Search on Modern CPUs and GPUs," *Proceedings of the 2010 ACM SIGMOD International Conference on Management of Data*, 2010. DOI: 10.1145/1807167.1807206. 6

S. Kim and H. Han, "Efficient SIMD Code Generation for Irregular Kernels," *Proceedings of the 17th ACM SIGPLAN Symposium on Principles and Practice of Parallel Programming*, 2012. DOI: 10.1145/2145816.2145824. 25

C. Kozyrakis and D. A. Patterson, "Scalable, Vector Processors for Embedded Systems," *IEEE Micro*, vol. 23, no. 6, pp. 36–45, 2003. DOI: 10.1109/MM.2003.1261385. 22

R. Krashinsky, C. Batten, M. Hampton, S. Gerding, B. Pharris, J. Casper, and K. Asanovic, "The Vector-Thread Architecture," *Proceedings of the 31st Annual International Symposium on Computer Architecture*, 2004. DOI: 10.1145/1028176.1006736. 21, 68

D. Kuck, Y. Muraoka, and S. C. Chen, "On the Number of Operations Simultaneously Executable in Fortran-like Programs and Their Resulting Speedup," *IEEE Transactions on Computers*, vol. C-21, no. 12, pp. 1293–1310, 1972. DOI: 10.1109/T-C.1972.223501. 23

S. Kumar, D. Kim, M. Smelyanskiy, Y.-K. Chen, J. Chhugani, C. J. Hughes, C. Kim, V. W. Lee, and A. D. Nguyen, "Atomic Vector Operations on Chip Multiprocessors," *Proceedings of the 35th Annual International Symposium on Computer Architecture*, 2008. DOI: 10.1145/1394608.1382154. 81

Y. Lee, R. Avizienis, A. Bishara, R. Xia, D. Lockhart, C. Batten, and K. Asanovic, "Exploring the Tradeoffs between Programmability and Efficiency in Data-Parallel Accelerators," *Proceedings of the 38th International Symposium on Computer Architecture*, 2011. DOI: 10.1145/2024723.2000080. 21

J. Leverich, H. Arakida, A. Solomatnikov, A. Firoozshahian, M. Horowitz, and C. Kozyrakis, "Comparing Memory Systems for Chip Multiprocessors," *Proceedings of the 34th International Symposium on Computer Architecture*, 2007. DOI: 10.1145/1273440.1250707. 45

A. Levinthal and T. Porter, "Chap - A SIMD Graphics Processor," *Proceedings of the 11th Annual Conference on Computer Graphics and Interactive Techniques*, 1984. DOI: 10.1145/964965.808581. 75

J. S. Liptay, "Structural Aspects of the System/360 Model 85, II: The Cache," *IBM Systems Journal*, vol. 7, no. 1, pp. 15-21, 1968. DOI: 10.1147/sj.71.0015. 63

Microsoft Corporation, http://msdn.microsoft.com/en-us/library/26td21ds.aspx. 23

K. Miura and K. Uchida, "FACOM Vector Processor System: VP-100/VP-200," *Proceedings of the NATO Advanced Research Workshop on High-Speed Computing*, 1983. 37, 48, 71

D. Nuzman, I. Rosen, and A. Zaks, "Auto-Vectorization of Interleaved Data for SIMD," *Proceedings of the 2006 ACM SIGPLAN Conference on Programming Language Design and Implementation*, 2006. DOI: 10.1145/1133981.1133997. 25

D. Nuzman and A. Zaks, "Outer-Loop Vectorization: Revisited for Short SIMD Architectures," *Proceedings of the 17th international Conference on Parallel Architectures and Compilation Techniques*, 2008. DOI: 10.1145/1454115.1454119. 24, 34

Nvidia Corporation, "CUDA C Programming Guide v6.5," http://docs.nvidia.com/cuda/cuda-c-programming-guide. 24, 60

Oracle Corporation, "Project Fortress," https://projectfortress.java.net, 2011. 24

D. Padua, R. Eigenmann, J. Hoeflinger, P. Petersen, P. Tu, S. Weatherford, and K. Faigin, "Polaris: A New-Generation Parallelizing Compiler for MPP's," Technical Report 1306, University of Illinois at Urbana-Champaign, Center for Supercomputing Research and Development, June 1993. 22

A. Pajuelo, A. Gonzalez, and M. Valero, "Speculative Dynamic Vectorization," *Proceedings of the 29th Annual International Symposium on Computer Architecture*, 2002. DOI: 10.1109/ISCA.2002.1003585. 81

S. Palacharla, N. P. Jouppi, and J. E. Smith, "Complexity-Effective Superscalar Processors," *Proceedings of the 24th Annual International Symposium on Computer Architecture*, 1997. DOI: 10.1145/384286.264201. 14

Y. Park, J. J. K. Park, H. Park, and S. Mahlke, "Libra: Tailoring SIMD Execution Using Heterogeneous Hardware and Dynamic Configurability," *Proceedings of the 2012 45th Annual IEEE/ACM International Symposium on Microarchitecture*, 2012. DOI: 10.1109/MICRO.2012.17. 21

J. Park, R. M. Yoo, D. S. Khudia, C. J. Hughes, and D. Kim, "Location-Aware Cache Management for Many-Core Processors with Deep Cache Hierarchy," *Proceedings of the International Conference for High Performance Computing, Networking, Storage, and Analysis*, 2013. DOI: 10.1145/2503210.2503224. 46

PassMark Software, CPU Mark, Single Thread Performance, http://www.cpubenchmark.net/singleThread.html. xi

S. J. Pennycook, C. J. Hughes, M. Smelyanskiy, and S. A. Jarvis, "Exploring SIMD for Molecular Dynamics, Using Intel Xeon Processors and Intel Xeon Phi Coprocessors," *Proceedings of the 27th IEEE International Symposium on Parallel and Distributed Processing*, 2013. DOI: 10.1109/IPDPS.2013.44. 3

M. Plotnikov, A. Naraikin, and C. J. Hughes, "Loop Vectorization Method and Apparatus," United States Patent Application no. 13/994549, filed 2012, published 2014. 79

R. Rahman, "Intel Xeon Phi Coprocessor Architecture and Tools: The Guide for Application Developers," Apress, 2012. 71

G. Ren, P. Wu, and D. Padua, "Optimizing Data Permutations for SIMD Devices," *Proceedings of the 2006 ACM SIGPLAN Conference on Programming Language Design and Implementation*, 2006. DOI: 10.1145/1133981.1133996. 25

M. Rhu and M. Erez, "Maximizing SIMD Resource Utilization in GPGPUs with SIMD Lane Permutation," *Proceedings of the 40th Annual International Symposium on Computer Architecture*, 2013. DOI: 10.1145/2485922.2485953. 75

M. Rhu, M. Sullivan, J. Leng, and M. Erez, "A Locality-Aware Memory Hierarchy for Energy-Efficient GPU Architectures," *Proceedings of the 46th Annual IEEE/ACM International Symposium on Microarchitecture*, 2013. DOI: 10.1145/2540708.2540717. 64

J. A. Rivers, G. S. Tyson, E. S. Davidson, and T. M. Austin, "On High-Bandwidth Data Cache Design for Multi-Issue Processors," *Proceedings of the 30th Annual IEEE/ACM International Symposium on Microarchitecture*, 1997. DOI: 10.1109/MICRO.1997.645796. 17

S. Rivoire, R. Schultz, T. Okuda, and C. Kozyrakis, "Vector Lane Threading," *Proceedings of the 2006 International Conference on Parallel Processing*, 2006. DOI: 10.1109/ICPP.2006.74. 21

T. G. Rogers, M. O'Connor, and Tor. M. Aamodt, "Divergence-Aware Warp Scheduling," *Proceedings of the 46th Annual IEEE/ACM International Symposium on Microarchitecture*, 2013. DOI: 10.1145/2540708.2540718. 49, 58

V. Saraswat, B. Bloom, I. Peshansky, O. Tardieu, and D. Grove, "X10 Language Specification," 2014. 24

N. Satish, C. Kim, J. Chhugani, A. D. Nguyen, V. W. Lee, D. Kim, and P. Dubey, "Fast Sort on CPUs and GPUs: A Case For Bandwidth Oblivious SIMD Sort," *Proceedings of the 2010 ACM SIGMOD International Conference on Management of Data*, 2010. DOI: 10.1145/1807167.1807207. 6

N. Satish, C. Kim, J. Chhugani, H. Saito, R. Krishnaiyer, M. Smelyanskiy, M. Girkar, and P. Dubey, "Can Traditional Programming Bridge the Ninja Performance Gap for Parallel Computing Applications?" *Proceedings of the 39th Annual International Symposium on Computer Architecture*, 2012. DOI: 10.1145/2366231.2337210. 24

J. Shin, "Introducing Control Flow into Vectorized Code," *Proceedings of the 16th International Conference on Parallel Architecture and Compilation Techniques*, 2007. DOI: 10.1109/PACT.2007.4336219. 25

J. Shin, M. W. Hall, and J. Chame, "Evaluating Compiler Technology for Control-Flow Optimizations for Multimedia Extension Architectures," *Proceedings of the 6th Workshop on Media and Streaming Processors*, 2004. DOI: 10.1016/j.micpro.2009.02.002. 41

M. Smelyanskiy, D. Holmes, J. Chhugani, A. Larson, D. M. Carmean, D. Hanson, P. Dubey, K. Augustine, D. Kim, A. Kyker, V. W. Lee, A. D. Nguyen, L. Seiler, and R. Robb, "Mapping High-Fidelity Volume Rendering for Medical Imaging to CPU, GPU, and Many-Core Architectures," *IEEE Transactions on Visualization and Computer Graphics*, vol. 15, no. 6, pp. 1563–1570, 2009. DOI: 10.1109/TVCG.2009.164. 6

M. Smelyanskiy, S. Kumar, D. Kim, J. Chhugani, C. Kim, C. J. Hughes, V. W. Lee, A. D. Nguyen, and Y.-K. Chen, "Vector Instructions to Enable Efficient Synchronization and Parallel Reduction Operations," United States Patent Application no. 12/079774, filed 2008, published 2009. 84

M. Smelyanskiy, J. Sewall, D. D. Kalamkar, N. Satish, P. Dubey, N., Astafiev, I. Burylov, A. Nikolaev, S. Maidonov, S. Li, S. Kulkarni, C. H. Finan, and E. Gonina, "Analysis and Optimization of Financial Analytics Benchmark on Modern Multi- and Many-Core IA-Based Architectures," *2012 SC Companion: High Performance Computing, Networking, Storage, and Analysis*, 2012. DOI: 10.1109/SC.Companion.2012.139. 6

B. J. Smith, "Architecture and Applications of the HEP Multiprocessor Computer System," *Proceedings of SPIE - Real-Time Signal Processing IV*, 1981. 12

J. E. Smith, G. Faanes, and R. Sugumar, "Vector Instruction Set Support for Conditional Operations," *Proceedings of the 27th Annual International Symposium on Computer Architecture*, 2000. DOI: 10.1145/342001.339693. 73

M. N. Stuttle, "A Gaussian Mixture Model Spectral Representation for Speech Recognition," Ph.D. Thesis, Cambridge University, Department of Engineering, July 2003. 4

X. Tian, H. Saito, M. Girkar, S. V. Preis, S. S. Kozhukhov, A. G. Cherkasov, C. Nelson, N. Panchenko, and R. Geva, "Compiling C/C++ SIMD Extensions for Function and Loop Vectorization on Multicore-SIMD Processors," *Proceedings of the 2012 IEEE 26th International Parallel and Distributed Processing Symposium Workshops and PhD Forum*, 2012. DOI: 10.1109/IPDPSW.2012.292. 25

K. Trifunovic, D. Nuzman, A. Cohen, A. Zaks, and I. Rosen, "Polyhedral-Model Guided Loop-Nest Auto-Vectorization," *Proceedings of the 18th International Conference on Parallel Architectures and Compilation Techniques*, 2009. 24, 34

D. M. Tullsen, S. J. Eggers, and H. M. Levy, "Simultaneous Multithreading: Maximizing On-Chip Parallelism," *Proceedings of the 22nd Annual International Symposium on Computer Architecture*, 1995. DOI: 10.1109/ISCA.1995.524578. 12, 16

A. S. Vaidya, A. Shayesteh, D. H. Woo, R. Saharoy, and M. Azimi, "SIMD Divergence Optimization Through Intra-Warp Compaction," *Proceedings of the 40th Annual International Symposium on Computer Architecture*, 2013. DOI: 10.1145/2485922.2485954. 75

S. Venkataramani, V. K. Chippa, S. T. Chakradhar, K. Roy, and A. Raghunathan, "Quality Programmable Vector Processors for Approximate Computing," *Proceedings of the 46th Annual IEEE/ACM International Symposium on Microarchitecture*, 2013. DOI: 10.1145/2540708.2540710. 31

T. Watanabe, "Architecture and Performance of NEC Supercomputer SX System," *Parallel Computing*, vol. 5, pp. 247–255, 1987. DOI: 10.1016/0167-8191(87)90021-4. 37, 48, 71

W. J. Watson, "The TI ASC — A Highly Modular and Flexible Super Computer Architecture," *Proceedings of the AFIPS 1972 Fall Joint Computer Conference*, 1972. DOI: 10.1145/1479992.1480022. 10

J. Wawrzynek, K. Asanovic, B. Kingsbury, J. Beck, D. Johnson, and N. Morgan, "SPERT-II: A Vector Microprocessor System," *IEEE Computer*, vol. 29, no. 3, pp. 79–86, 1996. DOI: 10.1109/2.485896. 64

P. Wu, A. E. Eichenberger, A. Wang, and P. Zhao, "An Integrated Simdization Framework Using Virtual Vectors," *Proceedings of the 19th Annual International Conference on Supercomputing*, 2005. DOI: 10.1145/1088149.1088172. 25, 45

Author's Biography

CHRISTOPHER J. HUGHES

Christopher J. Hughes is a principal engineer at Intel Labs, where he joined in August 2003. He received his Ph.D. in Computer Science from the University of Illinois at Urbana-Champaign in 2003, Master of Science degree in Computer Science from the University of Illinois at Urbana-Champaign in 2000, and Bachelor of Science degree in Electrical Engineering and Bachelor of Arts degree in Computer Science from Rice University in 1998. He led the teams that defined the gather instructions in Intel AVX2, the gather and scatter instructions in Intel AVX-512, and Intel's AVX-512CD instructions. His research focuses on highly parallel architectures for compute- and data-intensive applications.